Java语言项目化教程

微课视频版

◎ 徐舒 易凡 编著

清华大学出版社
北京

内 容 简 介

本书通过精心设计的"模拟电子屏"辅助读者快速完成"俄罗斯方块""贪吃蛇""飞机大战"等经典游戏的开发，并在完成游戏开发的过程中学习 Java 语言核心知识和面向对象的设计思想，让读者切身体会到程序设计的乐趣和魅力。全书共 14 章，分别介绍 Java 语言概述、Java 数据类型及运算符、控制程序设计、数组、面向对象基础、面向对象特性、集合与泛型、异常处理、字符串、输入输出、多线程、数据库编程、网络编程和综合应用等知识。

本书非常适合作为高等学校各类专业"Java 语言程序设计"课程的教材，也可以作为编程爱好者的自学辅导书。

本书封面贴有清华大学出版社防伪标签，无标签者不得销售。
版权所有，侵权必究。举报：010-62782989，beiqinquan@tup.tsinghua.edu.cn。

图书在版编目(CIP)数据

Java 语言项目化教程：微课视频版/徐舒，易凡编著．—北京：清华大学出版社，2023.3(2023.8重印)
(21 世纪新形态教·学·练一体化系列丛书)
ISBN 978-7-302-62680-0

Ⅰ.①J… Ⅱ.①徐… ②易… Ⅲ.①JAVA 语言－程序设计－高等学校－教材 Ⅳ.①TP312.8

中国国家版本馆 CIP 数据核字(2023)第 025964 号

责任编辑：陈景辉　李　燕
封面设计：刘　键
责任校对：韩天竹
责任印制：丛怀宇

出版发行：清华大学出版社
网　　址：http://www.tup.com.cn，http://www.wqbook.com
地　　址：北京清华大学学研大厦 A 座　　邮　编：100084
社 总 机：010-83470000　　邮　购：010-62786544
投稿与读者服务：010-62776969，c-service@tup.tsinghua.edu.cn
质量反馈：010-62772015，zhiliang@tup.tsinghua.edu.cn
课件下载：http://www.tup.com.cn，010-83470236

印 装 者：北京鑫海金澳胶印有限公司
经　　销：全国新华书店
开　　本：203mm×260mm　　印　张：15　　字　数：384 千字
版　　次：2023 年 4 月第 1 版　　印　次：2023 年 8 月第 2 次印刷
印　　数：1501～3000
定　　价：49.90 元

产品编号：096082-01

PREFACE 前言

Java语言是面向对象编程语言的代表,很好地体现了面向对象的理论,允许程序设计者以整体的思维方式进行程序设计。因其具有安全性、稳定性,拥有完善的多线程机制和强大的网络编程能力等特点,使其在面向对象和网络编程中占主导地位。尤其在互联网时代,Java语言大放异彩,成为常用的编程语言之一,因此学习Java语言非常有必要。

大多数传统的Java语言课程,将主要精力放在对Java语言语法细节的介绍上。学生们从一开始就陷入琐碎的细节之中,无法感受到程序设计的乐趣。这样的传统课程带给学生的学习体验常常比较糟糕,这会将许多本来热爱计算机的人拒之门外。本书是《C语言项目化教程(微课视频版)》的姊妹篇,两者设计思想一脉相承,都借鉴了斯坦福大学"编程方法学"的思想,通过"微项目"在简化的环境中介绍编程。本书设计的"模拟电子屏"微项目就像围棋一样,规则非常简单,但是变化万千。与C语言版本相比,Java语言版的微项目是采用面向对象的思想设计的,内容更加丰富多彩,能够实现更加复杂的游戏,充分展现了面向对象的魅力与价值。

读者可以通过"模拟电子屏"项目快速掌握Java的核心技术,迅速完成"俄罗斯方块""贪吃蛇""飞机大战"等经典游戏,并且通过这些游戏,找到彼此之间的联系,建立起通用的框架,然后就能快速编写各种经典小游戏,从而体会到作为一名设计者的乐趣,进而掌握程序设计最本质的内容。在熟练掌握Java核心技术后,可以不断升级小游戏,增加新的功能,使其成为一个复杂的网络游戏系统。在这个过程中读者可以学习到数据库和网络编程相关的知识,并且为学习如Spring等经典开源框架奠定良好的基础。

简而言之,读者在完成这些项目的过程中,一条经典的Java工程师进阶路线图就隐藏其中。这条路线就是从Java核心技术的学习到设计模式、流行开源框架的使用,然后是高并发、分布式系统的设计。这是一条精彩而有趣的路线,初学者可能对这条路线很陌生,但是随着学习的深入,会慢慢理解这条路线背后的逻辑。当然,这条路不是一蹴而就的,需要循序渐进。

本书受篇幅所限,不可能将这条路线中的所有内容包含其中,主要聚焦在Java核心技术的学习,并且在不断重构代码的过程中,会涉及部分经典的设计模式,如工厂模式、建造者模式等,让读者切身感受到设计模式的作用。读者在学习的时候,也需要不断重构代码,在重构的过程中,不断精进程序设计能力。

通过"模拟电子屏"项目学习Java语言,读者不仅可以轻松愉快地学习到Java语言的知识,而且能切身感受到,作为面向对象语言的代表,Java语言在解决程序开发中复杂性问题上的优势,正如Java语言设计者的初衷一样:Java语言的目标是减少开发稳健性代码所需的时间和困难。

本书主要内容

本书共分为 14 章,各章主要内容如下:

第一部分为基础篇,包括第 1~6 章。

第 1 章介绍程序及程序设计的基本概念,并且通过简单的案例介绍 Java 语言程序的基本结构和特点。

第 2 章介绍数据类型、运算规则,如何读取和操作数据。

第 3 章介绍 Java 的流程控制语句,包括选择控制结构语句和循环控制结构语句。

第 4 章介绍数组的定义、引用,以及数组的应用。

第 5、6 章介绍面向对象的编程思想、基本概念和特征,主要包括类的设计、对象的创建、抽象类和接口以及封装性、继承性和多态性。这是本书的重点内容,只有掌握了面向对象的编程思想,才能真正理解 Java 语言。

第二部分为提高篇,包括第 7~13 章。

第 7 章介绍集合与泛型,主要讲述集合和泛型的概念,以及常用集合的使用。

第 8~10 章介绍异常处理、字符串和输入输出常用的类,主要讲述异常处理机制、以及字符串类常用方法和流的概念,以及文件读写操作。

第 11 章介绍多线程,包括线程的创建、生命周期、调度方式以及线程同步、死锁等内容。

第 12 章介绍 JDBC 的基本概念,包括数据库访问步骤,以及如何使用 JDBC 对数据库进行操作。

第 13 章介绍网络编程的相关知识,包括网络通信协议以及使用 TCP 和 UDP 进行网络程序的编写。

第三部分为综合应用篇,包括第 14 章。

第 14 章介绍综合应用案例:网络版"飞机大战"游戏,利用多线程、数据库与网络编程等相关知识完成较为复杂的网络版游戏,并通过该案例掌握网络游戏编程的框架设计。

本书特色

(1) 引导读者使用面向对象的思维去思考问题和解决问题。"模拟电子屏"是一个小而美的游戏引擎,采用面向对象的思维方式设计完成,扩展性较强,通过它可以完成很多有趣的游戏。在完成较多游戏之后,读者会发现游戏之间的共性问题,逐步学会使用面向对象的思维去设计程序,提高程序的复用性和可扩展性。

(2) 以项目式游戏开发的实战案例,驱动 Java 语言编程的学习。本书的项目采用"小步快跑,快速迭代"的互联网产品设计方法,将一个功能简单的小项目,逐步迭代成一个复杂的系统。在此过程中,读者可以直观地感受到软件设计的魅力和乐趣,并从中学习到软件设计的思想和方法。

(3) 语言简洁、案例实用、体例清晰、配套资源丰富,对初学者友好。本书语言通俗易懂、简洁明了;对 Java 8~Java 19 中重要的新特性进行讲解,涉及了 Lambda 表达式和接口的新特性等内容;案例实用性强,符合企业用人实际需求;结构层次分明,各章相互关联、逐步递进,便于读者自学或高校选作教材使用。

配套资源

为了便于教与学,本书配有微课视频(700 分钟)、源代码、教学课件、教学大纲、教案、题库。

（1）微课视频获取方式：先刮开并用手机扫描本书封底的文泉云盘防盗码，再扫描书中相应的视频二维码，即可观看视频。

（2）源代码获取方式：先用手机扫描本书封底的文泉云盘防盗码，再扫描下方二维码即可获取。

源代码

（3）其他配套资源获取方式：用手机扫描本书封底的"书圈"二维码，关注后回复本书书号即可下载。

读者对象

本书既可作为高等学校"Java 语言程序设计"课程的教材，也可以作为编程爱好者的自学辅导书。读者对象包括零基础的初学者；具有一定基础，但是缺乏项目经验的读者；想从事 Java 语言软件开发相关工作的读者。

本书由徐舒工程师和易凡教授合作完成。徐舒曾是著名 IT 公司和法国国家科学中心 LIMOS 实验室的工程师，有着丰富的工程经验和教学经验。易凡是武汉大学物理学院教授，有着丰富的教学和科研经验。

本书在策划和出版的过程中，得到许多人的帮助，在此对这些人表示衷心的感谢。感谢作者的导师武汉大学孙洪教授的指导和帮助；感谢武汉理工大学刘岚教授的指导和帮助；感谢 LIMOS 实验室 Jean Connier 博士在作者于法国工作期间给予的帮助和支持；感谢张金龙、姚敏、余倩、王健、杨汉、吴俊、杨彬、于满洋、洪自华、陆奎良、李志龙等众多互联网公司的工程师们的支持和帮助。

在本书的编写过程中，参考了诸多相关资料，在此向文献资料的作者表示衷心的感谢。

限于编者水平及时间仓促，书中难免存在疏漏之处，欢迎读者批评指正。

编　者
2023 年 1 月

第一部分 基础篇

第 1 章 Java 语言概述 ... 3
1.1 Java 语言的特点 ... 3
1.2 编程环境 ... 4
1.2.1 Java 语言软件开发包 ... 4
1.2.2 集成开发环境介绍 ... 4
1.3 "模拟电子屏"项目介绍 ... 5
1.3.1 项目简介 ... 5
1.3.2 项目结构介绍 ... 5
1.3.3 项目核心类和方法介绍 ... 6
1.4 简单的 Java 语言程序示例 ... 7
1.5 注释 ... 9
1.6 综合案例:"俄罗斯方块"向下运动 ... 10
习题 ... 12

第 2 章 Java 数据类型及运算符 ... 13
2.1 数据类型 ... 13
2.2 变量和常量 ... 15
2.2.1 变量 ... 15
2.2.2 常量 ... 17
2.3 运算符与表达式 ... 18
2.3.1 赋值运算符与赋值表达式 ... 18
2.3.2 算术运算符与表达式 ... 19
2.2.3 关系运算符与关系表达式 ... 21
2.3.4 逻辑运算符与逻辑表达式 ... 21
2.3.5 逗号运算符与逗号表达式 ... 23
2.3.6 运算符优先级 ... 23

2.4 类型转换 ··· 23
　　2.4.1 自动类型转换 ··· 23
　　2.4.2 强制类型转换 ··· 24
2.5 综合案例："贪吃蛇"的运动 ··· 24
习题 ··· 26

第 3 章 控制程序设计 ··· 27

3.1 选择控制结构语句 ·· 27
　　3.1.1 if 语句 ·· 27
　　3.1.2 switch 语句 ·· 31
3.2 循环控制结构语句 ·· 34
　　3.2.1 while 语句 ·· 34
　　3.2.2 do-while 语句 ··· 35
　　3.2.3 for 语句 ··· 36
　　3.2.4 三种循环的比较 ·· 37
　　3.2.5 嵌套循环语句 ··· 37
　　3.2.6 break 语句和 continue 语句 ·· 38
3.3 综合案例：按键控制"贪吃蛇"运动 ·· 40
习题 ··· 41

第 4 章 数组 ·· 43

4.1 一维数组 ··· 43
　　4.1.1 一维数组的定义 ·· 43
　　4.1.2 一维数组的初始化 ·· 44
　　4.1.3 一维数组的使用 ·· 45
4.2 二维数组 ··· 47
　　4.2.1 二维数组的定义 ·· 48
　　4.2.2 二维数组的初始化 ·· 48
　　4.2.3 二维数组的引用 ·· 49
4.3 综合案例："贪吃蛇"游戏 ·· 50
习题 ··· 52

第 5 章 面向对象基础 ··· 54

5.1 面向对象概述 ·· 54
5.2 类和对象 ··· 55
　　5.2.1 对象的创建与使用 ·· 55
　　5.2.2 类的定义 ·· 56
　　5.2.3 访问控制符 ·· 61
　　5.2.4 方法的重载 ·· 62

5.2.5　构造方法 ··· 63
　　　5.2.6　static 关键字 ·· 67
　5.3　综合案例：重构"贪吃蛇"游戏 ··· 69
　习题 ·· 72

第 6 章　面向对象特性

　6.1　类的继承 ·· 73
　　　6.1.1　继承的概念 ··· 73
　　　6.1.2　方法重写 ··· 75
　　　6.1.3　super 关键字的使用 ··· 76
　　　6.1.4　子类的构造方法及调用过程 ·· 77
　　　6.1.5　final 修饰符 ·· 79
　　　6.1.6　Object 类 ·· 80
　6.2　抽象类和接口 ·· 81
　　　6.2.1　抽象类 ·· 81
　　　6.2.2　接口 ·· 83
　6.3　多态 ··· 90
　　　6.3.1　多态概述 ··· 90
　　　6.3.2　对象的类型转换 ·· 94
　　　6.3.3　接口实现多态 ··· 95
　6.4　内部类 ··· 97
　　　6.4.1　静态内部类 ··· 98
　　　6.4.2　非静态内部类 ··· 101
　6.5　综合案例："地图"编辑器 ·· 105
　习题 ··· 109

第二部分　提高篇

第 7 章　集合与泛型

　7.1　集合的概念 ·· 113
　7.2　Collection 接口与实现类 ·· 114
　　　7.2.1　List 接口与实现类 ·· 114
　　　7.2.2　Set 接口与实现类 ··· 118
　　　7.2.3　Collection 集合遍历 ··· 120
　7.3　Map 接口与实现类 ··· 122
　7.4　泛型 ··· 123
　7.5　综合案例："飞机大战"游戏 ··· 124
　习题 ··· 126

第 8 章 异常处理 ... 127

8.1 异常处理的方法 ... 127
8.1.1 异常的概念 ... 127
8.1.2 异常的捕获和处理 ... 129
8.1.3 异常的抛出 ... 130
8.1.4 自定义异常 ... 132
8.2 综合案例：重构"飞机大战"游戏 ... 133
习题 ... 135

第 9 章 字符串 ... 136

9.1 String 类 ... 136
9.1.1 创建 String 类对象 ... 136
9.1.2 字符串类常用方法 ... 137
9.2 StringBuffer 类和 StringBuilder 类 ... 138
9.3 综合案例：数据加密和解密 ... 139
习题 ... 140

第 10 章 输入输出 ... 141

10.1 流的概念 ... 141
10.2 字节流 ... 142
10.2.1 InputStream 类和 OutputStream 类 ... 142
10.2.2 字节流读写文件 ... 143
10.2.3 缓冲字节流读写文件 ... 143
10.3 字符流 ... 144
10.3.1 字符流读写文件 ... 145
10.3.2 字符缓冲流读写文件 ... 145
10.4 标准输入输出流 ... 146
10.5 对象序列化 ... 148
10.6 综合案例：游戏数据的存档和读取 ... 150
习题 ... 153

第 11 章 多线程 ... 154

11.1 线程的概念 ... 154
11.2 线程的创建 ... 155
11.2.1 继承 Thread 类实现多线程 ... 155
11.2.2 通过 Runnable 接口实现多线程 ... 155
11.3 线程的状态与调度 ... 157
11.3.1 线程的状态 ... 157

	11.3.2 线程的调度 ··· 158
11.4	线程同步与对象锁 ·· 162
	11.4.1 线程安全 ··· 162
	11.4.2 同步方法 ··· 164
	11.4.3 同步代码块 ··· 166
	11.4.4 同步锁 ··· 167
	11.4.5 死锁问题 ··· 170
	11.4.6 线程通信 ··· 174
11.5	综合案例：多线程技术重构"飞机大战"游戏 ········ 178
习题	··· 187

第 12 章 数据库编程 ··· 188

12.1	JDBC 概述 ·· 188
12.2	JDBC 使用步骤 ·· 189
	12.2.1 加载驱动程序 ·· 189
	12.2.2 建立连接对象 ·· 189
	12.2.3 创建语句对象 ·· 190
	12.2.4 获取 SQL 语句执行结果 ··························· 191
	12.2.5 关闭对象，释放资源 ································ 191
12.3	DAO 设计模式 ·· 196
12.4	综合案例：用户管理系统 ··································· 198
习题	··· 202

第 13 章 网络编程 ··· 203

13.1	网络通信概述 ··· 203
13.2	TCP 通信 ·· 204
13.3	UDP 通信 ··· 208
13.4	综合案例：网络版用户管理系统 ·························· 210
习题	··· 214

第三部分　综合应用篇

第 14 章 综合应用：网络版"飞机大战" ······················ 217

习题 ··· 225

参考文献 ·· 226

第一部分

基础篇

第1章

Java语言概述

Java语言是一种经典的面向对象的编程语言，广泛应用于大型企业级应用程序和移动设备应用程序开发中。Java语言极好地体现了面向对象的理论，允许程序设计者以整体的思维方式进行程序设计。在互联网时代，因其具有安全性、分布性、平台无关性等特点，Java语言一跃成为最流行的语言。

Java语言与C/C++语言有着紧密的联系，从C语言和C++语言中继承了许多成分，甚至可以将Java语言看成是类C语言发展和衍生的产物。如果读者有C语言的基础，学习Java语言的时候，会有一种非常熟悉的感觉，从而能够快速上手。

1.1 Java语言的特点

Java语言由Sun公司（现已被Oracle公司收购）于1995年推出。当时推出的目的是解决现代程序设计上的问题。Java语言起源于一项较大规模消费者电子产品软件发展项目，该项目是为了发展小型、可靠、可移植、分布式的嵌入式系统。项目负责人詹姆斯·高斯林（James Gosling），也就是Java语言之父，最初计划采用C++语言完成该项目，但是在实现的过程中遇到了一些问题。最初的问题主要集中在编译器技术方面，尚且能应付，可是后来遇到了更多的困难，发现最好的解决方式就是更换编程语言，于是Java语言应运而生。

Java语言是一种简单、面向对象、分布式、安全、稳健、多线程和平台独立的语言。

1. 简单

Java语言是一个纯粹的面向对象的程序设计语言，它在继承C++语言面向对象技术的核心时，一方面舍弃了C++语言中容易引起错误的指针、多重继承等特性，另一方面增加了自动垃圾收集功能，该功能不仅简化了程序设计工作，而且能大幅度减少错误数量。

2. 面向对象

面向对象是 Java 语言的核心，在 Java 语言中，万物皆为对象，组成客观世界的实体被抽象为数据和对数据的操作，并使用类将其封装成为一个整体。这种方法提高了软件的重用性、灵活性和扩展性。

3. 分布式

Java 语言提供丰富的网络支持技术，这使得在 Java 语言中比在 C 或 C++ 语言中更容易创建网络连接。Java 应用程序可以通过统一资源定位符（Uniform Resource Locator，URL）开启和存取对象，访问网络资源如同访问本地文件一样简单，因此 Java 语言非常适用于分布式场景。

4. 安全

Java 语言的设计目的是适用于网络/分布式运算环境。为此，Java 语言非常强调安全性，以确保建立的系统不易被侵入。

5. 稳健

Java 语言不支持指针操作，增加了自动"垃圾回收"功能，保障了 Java 程序的稳健性，目前许多第三方以及银行的交易系统等都是使用 Java 语言进行开发。

6. 多线程

多线程是程序能同时执行多个任务的能力，例如玩游戏的同时进行语音聊天。多线程机制使应用程序能够并行执行，并且同步机制保证对共享数据的正确操作，提高程序的执行效率。Java 语言内置了多线程控制，使用 Java 语言可以方便地实现多线程程序。

7. 平台独立

Java 虚拟机(Java Virtual Machine，JVM)是运行字节码的虚拟计算机。Java 程序并不是将源程序直接编译成机器码，而是编译成一种称为字节码的中间代码，这种中间代码只有在 Java 虚拟机上才能运行。由于虚拟机的作用，使用 Java 语言在一个平台编写好的程序不需要修改就能在各种平台上运行，因此 Java 语言具有平台独立性。

1.2 编程环境

1.2.1 Java 语言软件开发包

学习程序设计离不开上机实践。"工欲善其事，必先利其器"，学习程序设计第一步是选择合适的开发环境。Java 开发环境主要是指 Java 语言的软件开发包（Java Development Kit，JDK）。JDK 是 Java 的核心，包括了 Java 编译器、Java 运行工具、Java 打包工具、Java 文档生成工具。

开发 Java 程序之前，必须安装好 JDK。JDK 版本在不断升级，并且不同操作系统下的 JDK 也不同，所以在 Oracle 官网上下载 JDK 的时候，需要根据计算机的操作系统选择下载相应的 JDK 安装包。

1.2.2 集成开发环境介绍

除了 JDK 外，实际开发过程中，还需要安装集成开发环境（Integrated Development

Environment,IDE)来进行 Java 程序开发。

常用的 Java 语言 IDE 主要是 Eclipse 软件和 IntelliJ IDEA 软件,两款软件各有千秋。本书的代码均在 Windows 操作系统下,采用 Eclipse 软件运行。Eclipse 软件是免安装的,下载之后可以直接使用。需要注意的是 Eclipse 软件的版本要与 JDK 版本一致,否则 Eclipse 软件无法正常使用。

1.3 "模拟电子屏"项目介绍

1.3.1 项目简介

配置好开发环境,并成功导入 Chapter 1 至 Chapter 14 项目之后,就可以借助"模拟电子屏"项目进行 Java 语言学习。"模拟电子屏"是通过 Java 语言模拟生成的"电子屏",其界面如图 1.1 所示。

在"模拟电子屏"上可以添加各种有趣的游戏精灵(游戏角色),并且通过程序控制精灵进行各种行为事件,实现有趣的游戏。精灵的位置可以通过屏幕边上的行和列数字确定,屏幕左侧的数字代表行数,上面的数字代表列数。例如,在位于屏幕上第 7 行第 3 列的方块中添加游戏精灵"飞机",如图 1.2 所示。

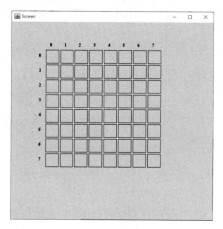

图 1.1 "模拟电子屏"界面

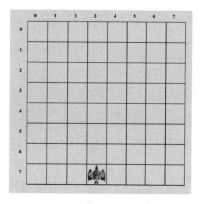

图 1.2 添加"飞机"的界面

通过"模拟电子屏"项目可以开发出各种各样的经典小游戏,从简单的方块类游戏"贪吃蛇""推箱子"到复杂精彩的网络版"飞机大战"游戏都能实现,如图 1.3 所示。

提示:把这块"模拟电子屏"想象成一块真正的"电子屏",充分发挥想象力和创造力,请试着思考通过它能完成什么任务。

1.3.2 项目结构介绍

为了辅助学习,本书提供 14 个项目,对应本书的 14 个章节。成功导入项目后,在 Eclipse 软件界面左侧的"包资源管理器"中会出现项目的树状结构菜单,如图 1.4 所示。

视频讲解

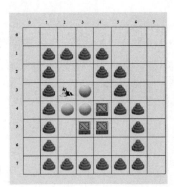
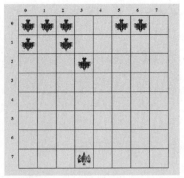

图 1.3　案例演示

从项目的树状结构菜单可知,项目以 Chapter 命名,如第 1 章对应项目的项目名为 Chapter1。每章的项目中包含着本章知识点案例和综合案例对应的代码,通过不同的包进行分隔,例如,例 1.1 对应的包名为"com.example1_1.screen",第 1 章的综合案例对应的包名为"com.example1_integrated.screen"。

通过项目名和包名很容易对应上每个章节中每个案例的代码。

1.3.3　项目核心类和方法介绍

所有的项目都是基于"模拟电子屏"展开的,通过在"模拟电子屏"上添加各种各样的游戏精灵,并且利用动画原理实现有趣的游戏。项目的核心类包括"模拟电子屏" Screen 类、方块 Cell 类和"游戏精灵"基类 GameObject 类。

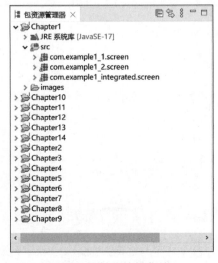

图 1.4　树状结构菜单

1. Screen 类

Screen 类是一个"模拟电子屏"类,该类提供了构造方法,可生成一个"模拟电子屏"实例对象,另外还提供了基本方法来控制"模拟电子屏"的使用,主要包括 add()、delay() 和 getKey() 方法,它们的作用如下所述。

(1) add() 方法。作用是将"游戏精灵"添加到屏幕上。例如向屏幕上添加方块,如图 1.5 所示。

(2) delay() 方法。作用是延时,可以实现许多有趣的动画和游戏。

(3) getKey() 方法。作用是获得键盘上正被按下的按键。通过该方法可以实现游戏中的按键操作功能,按下不同按键,执行不同操作,例如控制物体上下左右运动。

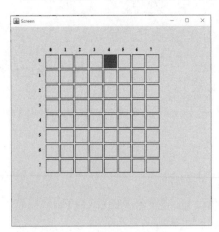

图 1.5　向屏幕上添加方块

2. Cell 类

Cell 类是一个方块类,该类提供了构造方法,可生成一个方块实例对象,并且提供了基本方法来控制方块的运动,分别是 moveUp()、moveDown()、moveLeft()、moveRight()和 moveTo()方法,它们的作用如下所述。

(1) moveUp()方法。作用是让方块向上运动一格。

(2) moveDown()方法。作用是让方块向下运动一格。

(3) moveLeft()方法。作用是让方块向左运动一格。

(4) moveRight()方法。作用是让方块向右运动一格。

(5) moveTo()方法。作用是让方块运动到某一个位置。

本书前 4 章主要学习 Java 语言的基础知识,为了避免增加学习的难度,所提供的"模拟电子屏"项目只能增加 Cell 类对象。

3. GameObject 类

GameObject 类是一个"游戏精灵"类,主要作用是创建各种游戏角色,并显示在"模拟电子屏"上。该类主要提供的方法包括设置或者获得"游戏精灵"在屏幕上的位置、生命值状态等。在学习了第 5、6 章类的基本概念和特性之后,读者可以通过继承 GameObject 类来创建各种有趣的游戏角色类,去完成自己感兴趣的游戏。

1.4 简单的 Java 语言程序示例

在熟悉了项目的基本结构和类的基本方法后,接下来就可以利用项目学习 Java 语言程序设计。双击项目树状结构菜单栏中的 Main.java 选项,如图 1.6 所示。

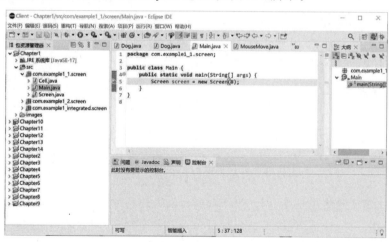

图 1.6 树状结构菜单栏中的 Main.java 选项

在右侧代码区就能阅读到第一个程序示例,具体代码如下:

```
public class Main {
    public static void main(String[ ] args) {
        Screen screen = new Screen(8);
    }
}
```

运行程序,出现一个8行8列的屏幕,如图1.7所示。

上述代码中主要包含两部分内容。

(1) 类定义。Java 语言是面向对象的语言,任何代码都必须放在一个类的定义中。"public class Main"的含义是定义一个名为"Main"的类。其中,"public"为类的访问修饰符,表示可以被任何成员访问,在后面的章节会详细讲解。"class"为关键字,表示定义的是一个类,"Main"为类名,其后一对大括号括起来的,称为类体。

(2) main()方法。main()方法是程序执行的起点,它类似于 C 语言的 main()函数。main()方法的格式如下:

```
public static void main(String[ ] args) {
}
```

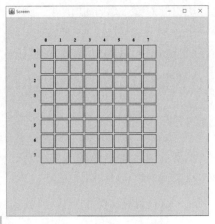

图1.7　8行8列屏幕图

main()方法将在第5章详细介绍,目前只需要了解所有的 Java 应用程序都是通过调用 main()方法开始执行的。在编写程序的时候,只需要按照上述格式完全复制即可。

在示例程序中,main()方法里编写了一条执行语句"Screen screen = new Screen(8);",它的作用是生成一个8行8列的 Screen 类的实例对象。类和对象是面向对象的基础概念,类是对一类事物的抽象描述,而对象表示现实中该类事物的个体。例如,猫类是一个抽象概念,而自己养的活生生的小花猫则是具体对象。Java 程序中可以使用 new 关键字来创建对象,具体语法格式如下:

```
类名 对象名 = new 类名();
```

语句"Screen screen = new Screen(8);"中的"new Screen(8)"用于创建 Screen 类的一个实例对象。修改括号里的数字,就会出现不同大小的屏幕。创建了对象之后,可以通过对象引用来访问对象中的所有成员,具体的格式如下:

```
对象引用.对象成员
```

例如:

```
Cell cell = new Cell(0,4);
screen.add(cell);
```

本段代码的作用就是将位于第0行、第4列位置的方块添加到 screen 屏幕上。

使用"对象引用.对象成员"这种方法非常符合人的思考模式,例如老师上课时常会讲到:张三同学来回答这个问题,或者李四同学的答案非常正确。张三、李四同学就是对象引用,而回答问题或者答案就是对象成员。

视频讲解

【例1.1】　将方块添加到屏幕上的某一个位置,如图1.8所示。

从图1.8中可知,屏幕的大小为8行8列,添加方块的位置为第5行第5列。设计程序时先按照基本框架写好 main()方法,接着生成 Screen 类的具体对象,然后生成方块对象,并将其添加到屏幕中,代码如下:

```
public class Main {
    public static void main(String[ ] args) {
        Screen screen = new Screen(8);
        Cell cell = new Cell(5,5);
        screen.add(cell);
    }
}
```

运行程序,屏幕中第5行第5列位置显示了一个方块。

【例1.2】 显示"俄罗斯方块"中的"T"形图形,如图1.9所示。

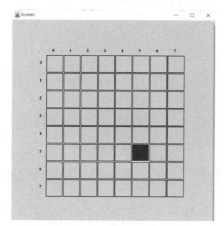
图1.8 添加方块

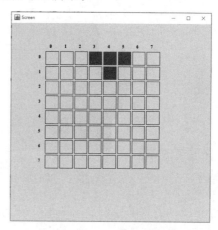
图1.9 "T"形"俄罗斯方块"

显示"T"形"俄罗斯方块"的形状,其实就是向屏幕上添加四个不同位置的方块,代码如下:

```
public class Main {
    public static void main(String[ ] args) {
        Screen screen = new Screen(8);
        Cell cell1 = new Cell(0,3);
        Cell cell2 = new Cell(0,4);
        Cell cell3 = new Cell(0,5);
        Cell cell4 = new Cell(1,4);

        screen.add(cell1);
        screen.add(cell2);
        screen.add(cell3);
        screen.add(cell4);
    }
}
```

运行程序,屏幕上显示出对应的图像。读者可以尝试显示"俄罗斯方块"游戏中其他形状的图像。

1.5 注释

"人人都能写出让计算机明白的程序,优秀的程序员能写出让人类理解的程序。"在编写程序的时候,为了让人们更加容易明白程序的含义和设计思想,通常会对代码进行解释和说明,这就是注释。注释只是为了提高代码的可读性,并不会被计算机编译,就像阅读文言文一样,为了帮助人们更好理解文章的意思,通常会在旁边添加各种注释,但是注释并不是原文章的内容。

在 Java 语言中有如下三种注释方法:

(1) 以"//"开始,以换行符结束的单行注释。

(2) 以"/*"开始、以"*/"结束的块注释,可以注释多行内容。

(3) 以"/**"开始、以"*/"结束的文档注释,也能注释多行内容,一般用在类、方法和变量前面,用来描述其作用。Java 语言的注释代码如下:

```
/**
 * 这是一个简单的Java语言程序
 * 作者:***
 * 时间:2021 年 12 月 30 日
 */
public class Main {
    public static void main(String[ ] args) {
        Screen screen = new Screen(8);        //生成8行8列大小的屏幕对象
        Cell cell = new Cell(5,5);             //生成方块对象
        screen.add(cell) ;                     //将方块添加到屏幕上
    }
}
```

虽然 Java 语言没有强制要求程序中一定要写注释,但是给代码写注释是一个良好的编程习惯。注释可以帮助他人和自己阅读代码,没有注释,即使自己写的代码,时间久了,阅读起来也会非常困难。因此,注释是程序非常重要的一部分。

视频讲解

1.6 综合案例:"俄罗斯方块"向下运动

Java 语言是面向对象的语言,其核心就是类的设计。如何将一个系统分解为多个类?这些类之间是什么关系?每一个类该如何设计?这些都是学习面向对象语言的重点内容,本节将带领读者在这些问题上取得突破。

例 1.2 虽然能够显示"俄罗斯方块"的形状,但是图像是静止的,如果想让方块运动起来,该如何实现?

由于计算机运行一条指令的速度太快,一般我们只能看到方块运动完之后的画面。如果要实现动画功能就需要设置延时程序,让方块的运动过程变慢。Cell 类提供了方块上下左右运动的方法,实现单个方块向下运动的代码如下:

```
public class Main {
    public static void main(String[ ] args) {
        Screen screen = new Screen(8);
        Cell cell = new Cell(0,4);
        screen.add(cell) ;
        screen.delay();              //延时一段时间
        cell.moveDown();             //方块向下运动
        screen.delay();
```

```
            cell.moveDown();
            screen.delay();
            cell.moveDown();
        }
}
```

实现了单个方块的运动,接下来实现"T"形"俄罗斯方块"的运动,如图1.10所示。

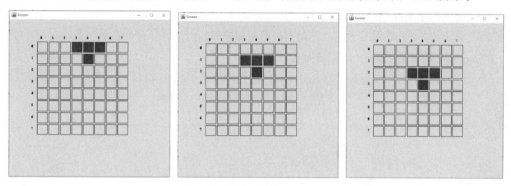

图1.10 "T"形"俄罗斯方块"向下运动示意图

实现"俄罗斯方块"的运动并不困难,只需实现组成"俄罗斯方块"的4个方块同时运动即可。例如,实现"俄罗斯方块"向下运动的代码如下:

```
public class Main {
    public static void main(String[ ] args) {
        Screen screen = new Screen(8);

        Cell cell1 = new Cell(0,3);                //创建4个方块对象
        Cell cell2 = new Cell(0,4);
        Cell cell3 = new Cell(0,5);
        Cell cell4 = new Cell(1,4);

        screen.add(cell1);                         //将4个方块添加到屏幕上
        screen.add(cell2);
        screen.add(cell3);
        screen.add(cell4);
        screen.delay();

        cell1.moveDown();                          //4个方块向下运动
        cell2.moveDown();
        cell3.moveDown();
        cell4.moveDown();

        screen.delay();
        cell1.moveDown();
        cell2.moveDown();
        cell3.moveDown();
        cell4.moveDown();

        screen.delay();
        cell1.moveDown();
```

```
            cell2.moveDown();
            cell3.moveDown();
            cell4.moveDown();
        }
    }
```

运行程序,屏幕上的"T"形"俄罗斯方块"从最上面开始不断向下运动。

上述代码虽然能实现功能,但是比较烦琐,如果想实现"俄罗斯方块"一直向下运动,直到屏幕底部,则需要继续编写大量重复代码。有没有更好的方法,让代码变得简单？在第3章会有办法解决这个问题,那就是使用循环语句简化程序。

习题

1.1 编写程序显示 16 行 16 列大小的屏幕。

1.2 编写程序显示"俄罗斯方块"的不同形状。

1.3 编写程序使"T"形"俄罗斯方块"向左运动。

第2章

Java数据类型及运算符

数据是程序的核心组成部分,程序中最常见的应用就是对数据进行处理。Java语言是强类型语言,数据的类型决定了其表现形式、存储方式和能进行的操作运算等。本章主要介绍Java语言的数据类型以及运算符的使用。

2.1 数据类型

Java语言是强类型语言,不同的数据类型表现形式、存储方式和能进行的操作运算会有所不同。这与生活中很多场景相似。例如,在预订机票时,可以选择经济舱或者商务舱等机票类型,不同类型的机票其价格以及服务是不一样的。程序中使用数据时,会根据数据类型分配所需要的存储空间,这种方法可以有效地利用存储空间,实现"量体裁衣"。

Java语言的数据类型分为两种:基本数据类型和引用数据类型。基本数据类型是Java语言内置的数据类型,引用数据类型是由程序设计者自己定义的数据类型。Java语言包含的数据类型如图2.1所示。

本章主要介绍基本数据类型,引用数据类型会在后面的章节中进行介绍。

Java语言的基本数据类型包括整数类型、浮点类型、字符类型和布尔类型。其中,整数类型按字节长度可以划分为4类,浮点类型按照精度可以划分为2类。

1. 整数类型

整数类型用来存储整数,根据不同的取值范围,Java语言提供了4种整数类型,包括字节型(byte)、短整型(short)、整型(int)和长整型(long)。4种类型所占存储空间以及取值范围如表2.1所示。

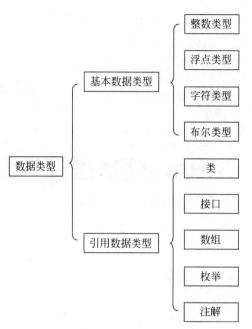

图 2.1 数据类型

表 2.1　4 种整数类型所占存储空间及取值范围

数据类型	描述	占用的存储空间/字节	取值范围	默认值
byte	字节型	1	-128~127	0
short	短整型	2	-32 768~32 717	0
int	整型	4	-2 147 483 648~2 147 483 647	0
long	长整型	8	-2^{63}~$2^{63}-1$	0L

提示：对于初学者来说,使用整数类型数据的时候,int 类型能够满足绝大部分程序要求。

2. 浮点类型

浮点类型用来存储小数,根据精度不同,分为单精度浮点型(float)和双精度浮点型(double)。表示浮点数时,双精度浮点型比单精度浮点型更精确,其中单精度浮点型精度是 8 位有效数字,双精度浮点型是 17 位有效数字,2 种类型所占存储空间以及取值范围如表 2.2 所示。

表 2.2　2 种浮点类型所占存储空间及取值范围

数据类型	描述	占用的存储空间/字节	取值范围	默认值
float	单精度浮点型	4	3.4e-38~3.4e+38	0.0f
double	双精度浮点型	8	1.79e-308~-1.79e+308	0.0d

其中,e 表示 10 为底的指数,3.4e+38 表示 $3.4×10^{38}$。

3. 字符类型

除了整数类型和浮点类型数据外,程序中有时候还需要用到字符类型数据,例如,成绩等级分为 A~E,游戏中按下的按键分别是"A""W""S""D"等,保存这些数据信息时,都需要字符型(char)数据。char 类型会占用 2 字节,本质上也是整数类型的一种,因为 char 类型实际上也是存储的整数。计算机使用数字编码来表示字符,即用特定的整数表示特定的字符。

4. 布尔类型

布尔类型(boolean)用于表示逻辑"真"和"假",对应的关键字分别为"true"和"false"。布尔类型的变量或表达式取值只能是"true"或者"false"。

2.2 变量和常量

2.2.1 变量

1. 变量的含义及基本操作

程序中,数据有两种表示形式:常量和变量。程序运行过程中,数值不发生变化的量是常量,发生变化的是变量。变量的作用是存储数据,存储的值时常会发生变化,所以被形象地称之为变量。变量是数据存储的基本概念,在程序中可以将变量理解成存储数据的容器。例如,使用计算机计算不同半径的圆的面积:

```
s = 3.14 * r * r;
```

这条语句中,有 s 和 r 两个变量,变量 r 存储的是圆的半径,变量 s 存储的是圆的面积。当变量 r 里面存储的值发生变化的时候,变量 s 的值也会随之变化,这样就能计算出不同半径的圆的面积。

通过上述例子可知,对变量的基本操作包括两部分:

(1) 向变量中存储数据。这个过程称为给变量"赋值",使用的符号为"=",其作用就是将数据放进容器之中。

(2) 获取变量的当前值。这个过程称为"取值",可以直接通过变量名获取变量的值。

如果把数据想象成实物,它的存取过程与生活中使用容器存储物品并无二致。初学者在使用变量的时候,可以将变量想象成存储数据的容器。给变量赋值,就是将数据存入容器之中。例如:

```
r = 1.0;
```

就是向变量 r 里存放数值 1.0;

```
s = 3.14 * r * r;
```

就是将计算好的圆面积的值存放在变量 s 之中。如果此时变量 r 的值为 1.0,则变量 s 的值为 3.14。

程序中,每一个变量都必须有一个名字作为标识,变量名代表内存中的存储单元。上述例子中的"s""r"就是变量名,存储着不同作用的数据。给变量命名的自由度较大,但是一般会根据变量的作用选择合适的名字,使其尽量有具体相关的含义,做到"顾名思义",例如:

```
price = 2.5;
```

通过名字就知道变量 price 存储的是价格。

2. 变量的命名规则

给变量命名除了尽量做到"顾名思义",还需要遵循一定的规则,Java语言命名详细规则如下:

(1) 只能由字母、数字、下画线(_)和美元符号($)组成,但是首字母不能是数字。例如,a、a9、_a、A_num、A$Num 为合法的名字;而 9a、a*num 为非法的名字。

(2) Java 语言是区分英文字母大小写的,即变量 A 和变量 a 是两个不同的变量。

(3) 变量名不能与关键字相同。关键字是具有特殊含义的字符串,通常也称为保留字,例如 int、float 等,这些关键字会在后续章节逐步学习。

3. 变量的定义

在 Java 语言中使用变量,必须先定义后使用。例如,在使用变量 a 之前,需要定义它。变量的定义包括变量的声明和赋值,变量的声明一般格式如下:

```
变量类型 变量名;
```

例如:

```
int a;
```

表示声明了一个新变量,变量名为 a,它的数据类型为 int,也就是整型。声明变量的同时,还可以对该变量进行初始化赋值,例如:

```
int a = 3;
```

表示变量 a 的初始值为 3。

变量定义成功之后,在编译时就能根据变量类型分配对应的存储空间。变量的定义就像预订机票一样,预定成功之后,就会将相应的座位预留出来。

变量先定义后使用,有一个好处就是避免错误。例如,变量 varValue 在使用的时候,如果变量名误写成 vaValue,系统会提示变量书写错误。如果没有先定义后使用这个规则,系统会将其当作一个新的变量进行操作,从而产生错误,并且这样的错误不容易查找出来,所以 Java 使用变量一定要先定义后使用。

视频讲解

【例 2.1】 编写程序,向屏幕添加一个方块,方块的位置随机。

由于方块的位置是随机的,所以需要定义两个整型变量 row 和 col 去存储方块的行和列位置信息,Screen 类里提供了产生随机整数的方法:getRandom()方法。使用该方法可以实现方块的位置随机,代码如下:

```java
public class Main {
    public static void main(String[ ] args) {
        Screen screen = new Screen(8);
        int row = screen.getRandom(8);        //随机产生 0~7 的整数
        int col = screen.getRandom(8);
        Cell cell = new Cell(row,col);
        screen.add(cell);
    }
}
```

运行程序,方块会随机出现在屏幕上。

2.2.2 常量

1. 常量的类型

常量就是程序运行过程中,数值不会发生变化的量。在Java语言中,常量包括整型常量、浮点型常量、字符常量、字符串常量、布尔常量。

1) 整型常量

整型常量是整数类型的数据,有二进制、八进制、十进制和十六进制4种表示方法。进制就是进位计数方法,对于任何一种进制,就表示每一位上的数运算都是逢多少进一位。例如,十进制是逢十进一,二进制就是逢二进一。以此类推,×进制,就是逢×进一。

在Java语言中,整型数据的不同进制表示方法如下:

(1) 二进制。数值以0b或者0B开头,每一位数字只能是0或者1,如0b00010001、0B00010001。

(2) 八进制。数值以0开头,每一位数字的范围为0~7,如0123。

(3) 十进制。数值没有任何前缀,每一位数字的范围为0~9,如123。

(4) 十六进制。数值以0x或者0X开头,每一位数字范围为0~9、A~F,如0x1AD、0XB0F。

2) 浮点型常量

浮点型常量是小数类型数据,比如圆周率π的值约为3.14,3.14就是浮点型常量。计算机在存储小数的时候采用了与整型数据不一致的方式,浮点型数据主要分为单精度、双精度两种类型。单精度浮点数后面以F或者f结尾,双精度浮点数以D或者d结尾,如:3.14f、9.8F、3.1415926D。

使用浮点型数据时也可以不加后缀,系统会将其默认为双精度浮点型,如3.14、9.8。

浮点型常量还可以通过指数形式表示,如:3.14e+10、9.5e-3。

3) 字符常量

字符常量用于表示单个字符,使用时需要用单引号括起来。例如,成绩等级'A'、'B'、'C'、'D'、'E'。

4) 字符串常量

字符串常量用于表示一串字符,使用时需要用一对双引号括起来。例如,"I am Chinese!"。

需要注意的是"A"和'A'是不同的,前者表示字符串,后者表示字符。当字符串里面不包含任何字符时,它是一个空字符串,长度为0。

5) 布尔常量

布尔常量就是布尔型数据,表示逻辑"真"和"假",只有"true"和"false"两个值,用来判断一个条件为"真"或者"假"。

2. 常量的定义

在Java语言中声明常量的格式如下:

```
final 常量类型 常量名;
```

定义常量的方式与定义变量的方式非常相似,唯一不同的就是常量类型前面加了关键字final,表示这是不可改变的量。例如:

```
final float PI = 3.14;
```

需要注意的是,根据常量命名规范,常量名一般需要大写。

视频讲解

【例 2.2】 编写程序,生成固定大小的屏幕,并向屏幕添加一个方块,方块的位置随机。

由于屏幕的大小在程序运行过程中不会发生变化,所以将其设置为常量,代码如下:

```
public class Main {
    public static void main(String[ ] args) {
        final int SIZE = 8;                    //屏幕大小设置为常量
        Screen screen = new Screen(SIZE);
        int row = screen.getRandom(SIZE);
        int col = screen.getRandom(SIZE);
        Cell cell = new Cell(row,col);
        screen.add(cell);
    }
}
```

运行程序,方块出现在屏幕随机位置上。

2.3 运算符与表达式

程序中除了存储数据外,还需要处理和计算数据。为了方便处理和计算各种数据,Java 语言提供了大量的运算符(又称操作符),可以在程序中进行算术运算、关系运算和逻辑运算等。

运算符是表示各种不同运算的符号,参与运算的各种数据被称为数据对象(也称作操作数)。表达式是由运算符将运算对象连接起来,符合 Java 语言规范的式子,每个表达式经过运算后都会有一个确定的值。

2.3.1 赋值运算符与赋值表达式

1. 赋值运算符

在 Java 语言中,赋值运算符"=",意思是将右侧的值赋给左侧的变量,例如:

```
int a = 3;
```

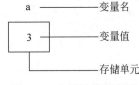

图 2.2 变量赋值示意图

语句的意思是将数值 3 赋给变量 a。也可以理解为,将数值 3 存储在变量 a 中,如图 2.2 所示。

赋值运算符"="与数学符号"等于"号意义并不一样,初学者容易混淆。以表达式"i=i+1"为例,在数学中这个表达式不能成立,因为一个数不可能等于它自身加 1。但是在计算机编程语言中这个表达式就能成立,其意义就是把变量 i 里的值加 1,重新赋值给变量 i。例如,变量 i 最初的值为 5,执行"i=i+1"之后,变量 i 的值就变成了 6。赋值的过程就是将数据存入内存存储单元的过程。

2. 复合赋值运算符

在赋值运算符"＝"之前加上其他运算符,就可以构成复合赋值运算符。例如:

```
sum + = i;
i + = 1;
```

在"＝"前加一个"＋"运算符就成了复合赋值运算符"＋＝"。"sum＋＝i"等价于"sum＝sum＋i","i＋＝1"等价于"i＝i+1"。

同理"i－＝1"等价于"i＝i－1","i＊＝2"等价于"i＝i＊2","i/＝2"等价于"i＝i/2","i％＝2"等价于"i＝i％2"。

并非一定要使用这些复合赋值运算符,但是使用它们可以简化程序,使代码变得更加紧凑。例如,计算衣服类商品打折之后的价格,代码为:

```
clothesPrice = clothesPrice * discount;
```

使用复合赋值运算符,可以修改为:

```
clothesPrice * = discount;
```

其中变量 clothesPrice 存储衣服价格,变量 discount 存储折扣信息。

3. 赋值表达式

赋值表达式就是由赋值运算符将一个变量和一个表达式连接起来的式子。它的一般格式为:

```
变量名 = 表达式
```

对赋值表达式求解过程为:先计算赋值运算符右边"表达式"的值,然后将计算结果赋值给运算符左边的变量。赋值表达式的值就是变量的值。例如:

```
a = 3 * 10
```

先计算表达式 3＊10 的值,结果为 30,然后将 30 赋给变量 a,变量 a 的值为 30,赋值表达式的值就是变量 a 的值。

2.3.2 算术运算符与表达式

1. 基本的算术运算符

Java 语言中基本的算术运算符有"＋""－""＊""/""％"5 种。

(1) ＋。加法运算符,使其两侧的值相加。
(2) －。减法运算符,使其左侧的数减去右侧的数。
(3) ＊。乘法运算符,使其两侧的数相乘。
(4) /。除法运算符,使其左侧的数除以右侧的数。
(5) ％。求模运算符,也称求余运算符,用于获取左侧的整数除以右侧的整数得到的余数。％两侧的数要求都为整数。

在计算机中使用"＋""－""＊""/"运算符跟数学中常用的四则运算规则非常相似,还可以带上括号。算术运算虽然简单,但是需要注意的是数学中使用乘法时有些情况下可以将乘号省略,如"3x"实际上是"3 * x",但在 Java 语言编程中"3 * x"不能省略中间的" * "号,否则编译器无法编译成功。算术运算中,利用求余运算,可以巧妙解决很多问题。例如,判断一个整数的奇偶性,可以利用对 2 求余,通过余数就可以确定整数的奇偶性。

2. 自增、自减运算符

自增、自减运算的作用就是将这个变量的值增加 1 或者减少 1,自增、自减符号分别为:"++""--"。例如,在让"俄罗斯方块"运动的代码中:

```
row = row + 1;
```

可以替换成

```
row++;
```

"row ++"与"row=row+1"效果相同,"row --"与"row=row-1"效果相同。

自增、自减运算符分为前缀、后缀方式。运算符在变量前面,称为前缀方式,表示变量在使用前自动加 1 或者减 1,如"++i""--i"。运算符在变量后面,称为后缀方式,表示变量在使用后自动加 1 或者减 1,如"i++""i--"。

"i++"与"++i"单独使用的时候,两者结果没有差别,变量 i 的值都增加了 1。但是当自增、自减运算符与其他符号连在一起使用时候,符号放的位置会对结果产生影响。例如:"y=++i"与"y=i++",i 的值都是增加了 1,但是变量 y 的值却是不一样。

"y=++i"等价于:

```
i = i + 1;
y = i;
```

先执行"i=i+1"然后执行"y=i"。假设运算前变量 i 的值为 5,则运算结束后变量 y 与变量 i 的值相同,都是 6。

而"y=i++"等价于:

```
y = i;
i = i + 1;
```

先执行"y=i"然后执行"i=i+1"。假设运算前变量 i 的值为 5,则运算结束后变量 y 的值为 5,变量 i 的值为 6,变量 i 的值比 y 的值大 1。

通过上述例子,可知自增、自减运算中,使用前缀方式和后缀方式,两者的结果有差别。对于经验丰富的程序员,使用自增或者自减运算符在某些应用场景下会写出非常紧凑、简洁的代码。但是这种方法降低了代码的可读性,并且容易产生计数错误,对于初学者来说需要慎用这种方式,不要将自增、自减运算符与其他运算符混用。编写程序的宗旨是易读易懂,并且不容易出错。

3. 算术表达式

算术表达式就是由算术运算符和括号将运算对象连接起来的式子,如"3.14 * r * r""m * c * c"。

每一个表达式经过运算都有确定的值。要获得表达式的值,就需要了解运算符之间的优先级,优先级决定了运算次序。算术运算符的优先级与数学的四则运算一致:括号优先级最高,接着

乘除,最后加减。另外,赋值运算符比算术运算符优先级低,例如:

```
s = 3.14 * r * r;
```

计算完算术表达式的值,得出的结果赋值到变量s中,所以变量s保存着运算结果。

2.2.3 关系运算符与关系表达式

1. 关系运算符

关系运算符的作用就是对两个数据进行比较,确定两个数据之间是否存在某种关系。Java语言提供了6种关系运算符,如表2.3所示。

表2.3 关系运算符及其含义

运算符	含义
==	等于
!=	不等于
>	大于
<	小于
>=	大于或等于
<=	小于或等于

关系运算非常简单,与数学中的关系运算很相似,区别在于符号上,"等于"在Java语言中是"==",因为"="已经被用作赋值符号。在编写程序的时候,"=="与"="容易混淆,导致程序错误,例如:

```
a = 5;
b == 5;
```

两者的意思不一样,前者是将5赋值给变量a,也就是变量a的值为5,而后者判断变量b的值与5是否相等。对于初学者,经常在判断相等关系的时候误写成赋值号,使用时需要注意。

2. 关系表达式

简单关系表达式是用关系运算符将运算对象组成的式子。例如,"a+b>=a*b""a!='A'"。

关系表达式的结果只有两种可能,真或者假。当关系成立的时候,结果为真,否则为假。也就是布尔类型的值"true"和"false"。例如,关系表达式"5>3"为真,表达式的值为"true"。关系表达式"2==3"为假,表达式的值为"false"。

关系表达式常用作判断条件,在选择语句或者循环语句中使用。例如,通过按键控制方块的运动中,需要使用关系表达式作为条件,根据按键的键值,执行相应的操作。

2.3.4 逻辑运算符与逻辑表达式

1. 逻辑运算符

逻辑运算符的作用是将多种关系表达式组合成更复杂的表达式,Java语言提供了3种逻辑运算符,如表2.4所示。

表 2.4 逻辑运算符及其含义

运算符	含义
&&	与
||	或
!	非

逻辑运算的结果也只有两种,真(true)或者假(false)。逻辑运算的规则如下:

(1) 逻辑"与"运算。只有"与"运算符两边的运算对象均为真的情况下才能为真,如表 2.5 所示。

(2) 逻辑"或"运算。只要"或"运算符两边至少有一个运算对象为真,运算结果为真,如表 2.6 所示。

(3) 逻辑"非"运算。原运算对象为真,进行"非"运算则结果为假,原运算对象为假,进行"非"运算则结果为真。简单理解就是非真即为假,非假即为真,如表 2.7 所示。

表 2.5 逻辑"与"运算的结果

变量 a	变量 b	结　果
true	true	true
true	false	false
false	true	false
false	false	false

表 2.6 逻辑"或"运算的结果

变量 a	变量 b	结　果
true	true	true
true	false	true
false	true	true
false	false	false

表 2.7 逻辑"非"运算的结果

变量 a	结　果
true	false
false	true

逻辑运算非常简单,但是用途非常广泛,在计算机应用中无处不在。例如,常用的购物网站,用户很轻松就能找到心仪的产品,背后的原理就是逻辑运算。计算机通过用户选择的条件进行逻辑运算,就能将用户中意的商品呈现出来。当一个条件满足的时候,则值为 true,否则值为 false,无论多么复杂的搜索条件最后都会转换成简单的逻辑运算。至繁归于至简,这也是计算机的魅力之一。

2. 逻辑表达式

逻辑表达式就是用逻辑运算符将关系表达式连接起来的式子。逻辑表达式的结果也只有"真"或"假"。例如,表达式"(5>2) && (10<6)"结果为假,因为只有一个运算对象为真;表达式"(5>2) || (10<6)"结果为真,因为有一个运算对象为真;表达式"!(5>2)"结果为假,因为运算对象为真,非真即为假。

使用逻辑运算符的时候要注意一种情况,例如,屏幕的大小为 8 行 8 列,判断方块的位置是否在屏幕内,也就是行与列的值范围为0～7,表达式为:"row＞＝0 && row＜8",而不是数学上的表达式"0＜＝row＜8"。

2.3.5 逗号运算符与逗号表达式

逗号运算符是一种特殊的运算符,作用是将多个表达式连接起来,一般格式为:

> 表达式1,表达式2,…,表达式n

逗号表达式求解过程为:先求解表达式1,再求解表达式2,从左到右依次按照顺序求值。整个逗号表达式的值为表达式n的值。例如:"a=(x=1,x+1)",其求解过程为:首先将1赋给变量x,然后执行x+1的计算,最后将整个结果赋给变量a,则变量a的值为2。

2.3.6 运算符优先级

一个复杂的表达式中可能有算术运算符、关系运算符等多种运算符,计算表达式值的时候,需要确定各种运算符的优先级。运算符优先级的一般规则为:逻辑"非"＞算术运算符＞关系运算符＞逻辑"与"＞逻辑"或"＞赋值运算符。

在编写程序的时候,尽量避免太复杂的表达式,当表达式过于复杂的时候,应该增加括号,消除歧义。例如:表达式"row＞0 && row＜8"与表达式"(row＞0) && (row＜8)"的运算结果虽然一致,但是加上括号可以使代码更加清晰。

当表达式复杂的时候,可以引入解释性变量,将表达式分解成比较容易理解的形式。

2.4 类型转换

在表达式运算中,有时候会遇到不同类型的数据进行混合运算。例如,某件商品的单价是20.5元,共买了10件,那么总价就是单价×数量,20.5是实型数据,10是整型数据,那么表达式"20.5 * 10"的结果是什么数据类型?

对于这种不同类型的数据混合运算的情况,需要进行数据类型转换。基本数据类型转换包括自动类型转换和强制类型转换。

2.4.1 自动类型转换

自动类型转换由系统自动完成,Java语言进行类型转换的一般规则如下。

(1) 低级别类型向高级别类型转换,类型的级别由低到高的顺序为:"byte、short、int、long、float、double"或者"char、int、long、float、double"。

例如,运算中存在整型与浮点型数据混合运算时,因为浮点型数据级别高于整型,所以需要将整型数据转换成浮点,则表达式"20.5 * 10"最终结果是实型数据"205.0",而不是整型数据"205"。

(2) 赋值运算的结果,以赋值运算符左边变量的类型为准,例如:

```
int count = 3.15 * 10;
```

执行后,变量 count 的结果为 31,而不是 31.5。虽然表达式"3.15 * 10"的结果是实型,但是进行赋值运算的时候,运算结果以左边变量的类型为准,变量 count 的数据类型是整型,所以变量 count 最终结果是 31。

通过这个例子会发现进行数据转换时,可能会导致数据信息丢失。

2.4.2 强制类型转换

如果需要将高级别的数据类型转换成低级别的数据类型,例如将 float 型转换成 int 型数据,则需要通过强制类型转换实现。强制转换的一般格式如下:

```
(类型名)(表达式)
```

在圆括号中给出希望转换的目标类型,随后是需要转换的表达式,例如:

```
int a = (int) 3.15 * 10;
```

运算的过程是,先将 3.15 转换成整型数据 3,然后计算表达式的值并赋值给变量 a,所以最终变量 a 的值为 30。

自动转换和强制转换都可能带来数据降级,从而导致信息丢失。因此在使用类型转换的时候,要小心谨慎地选择合适的类型,减少不必要的类型转换。

视频讲解

2.5 综合案例:"贪吃蛇"的运动

"贪吃蛇"游戏是一款经典的游戏,玩法非常简单,玩家使用方向键控制一条长长的"蛇"上下左右运动吃到"食物"。编写程序,实现长度为 3 的"贪吃蛇"向下运动,如图 2.3 所示。

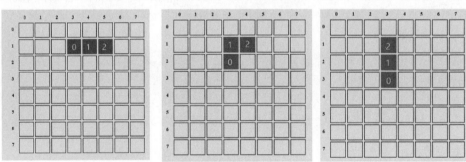

图 2.3 "贪吃蛇"向下运动示意图

"贪吃蛇"与"俄罗斯方块"的运动规则不同,"俄罗斯方块"是整体运动,每个方块运动方向都一致。例如,执行向左运动时,所有方块都相应左移一格。而"贪吃蛇"则不同,例如,"贪吃蛇"执行向下运动时,从图 2.3 可以看出,0 号位置的方块往下运动,而其他位置的方块运动方向却不相同,1 号和 2 号方块向左运动。每个方块运动方向好像没有规律,但是如果将"贪吃蛇"分为两部分:"蛇头"和"蛇身",就能发现规律。例如,0号位置的方块为"蛇头",其余位置的方块为"蛇身",那么,"贪吃蛇"运动的规律为:

(1)"蛇头"作为一个独立的部分,会根据按键控制的方向运动到相应的位置,比如向下运动。

(2)"蛇身"每一部分都是移动到前一块"蛇身"的位置。例如,2号方块移动到1号方块原来所在的位置,1号方块移动到0号方块原来所在的位置。

"贪吃蛇"运动代码如下:

```java
public class Main {
    public static void main(String[ ] args) {
        final int SIZE = 8;
        Screen screen = new Screen(SIZE );

        Cell head = new Cell(1,3);           //贪吃蛇的蛇头
        Cell body1 = new Cell(1,4);          //贪吃蛇的身体
        Cell body2 = new Cell(1,5);

        screen.add(head);                    //将方块添加到屏幕上
        screen.add(body1);
        screen.add(body2);

        screen.delay();
        head.moveDown();                     //蛇头向下运动
        body1.moveTo(head);                  //蛇身1运动到蛇头位置
        body2.moveTo(body1);                 //蛇身2运动到蛇身1位置
    }
}
```

运行程序,屏幕上的三个方块变成一个方块。简单分析就能找到问题,即所有的方块都移到"蛇头"的位置。解决方法很简单,就是将运动代码反过来,从"蛇尾"到"蛇头"依次运动,代码如下:

```java
public class Main {
    public static void main(String[ ] args) {
        final int SIZE = 8;
        Screen screen = new Screen(SIZE);
        Cell head = new Cell(1,3);           //贪吃蛇的蛇头
        Cell body1 = new Cell(1,4);          //贪吃蛇的身体
        Cell body2 = new Cell(1,5);

        screen.add(head);                    //将方块添加到屏幕上
        screen.add(body1);
        screen.add(body2);
        screen.delay();

        body2.moveTo(body1);                 //蛇身运动到前一个位置
        body1.moveTo(head);
        head.moveDown();                     //蛇头向下运动

        screen.delay();
```

```
        body2.moveTo(body1);
        body1.moveTo(head);
        head.moveDown();
    }
}
```

运行程序,"贪吃蛇"向下运动。上述代码虽然完成了任务,但是存在一个问题:如果"贪吃蛇"的长度较大,使用上述方法代码就非常烦琐,并且容易出现错误。在第 4 章,将学习数组,可以使代码变得简单且灵活。

习题

2.1 下面能用作 Java 语言标识符的是_____。
 A. 3mk B. a.f C. float D. ATM

2.2 若定义了"int x;"则将变量 x 强制转化成单精度类型的正确方法为_____。
 A. float(x) B. (float) x C. float x D. double(x)

2.3 已知定义了"int a; float b; double c;",执行语句"b=a+c;",变量 b 的数据类型为_____。
 A. int B. float C. double D. 不确定

2.4 能正确表示变量 x 的取值在 0~100 的表达式为_____。
 A. 0<=x<=100 B. x>=0 && x<=100
 C. x>=0 D. x<=100

2.5 能正确表示变量 x 的取值在 0~50 或 −100~−50 的表达式为_____。
 A. (0<=x<=50) || (−100<=x<=−50)
 B. (x>=0 && x<=50) && (x>=−100 && x<=−50)
 C. (x>=0 && x<=50) || (x>=−100 && x<=−50)
 D. (x>=0 || x<=50) && (x>=−100 || x<=−50)

2.6 已知定义了"int a;",执行语句"a=5/2;",变量 a 的结果是_____。

2.7 已知定义了"int a=5;",执行语句"a=a%3;",变量 a 的结果是_____。

2.8 编写程序,实现"方块"沿着屏幕对角线的运动。其中,屏幕的大小为 8 行 8 列。

第3章

控制程序设计

在前面两章,所有的程序都是顺序结构,即程序从 main()方法开始进入,然后逐条指令开始执行,直到所有指令都执行完。但是,就像生活一样,不是所有的事情都能按照计划执行,许多事情经常会随着条件的变化面临着各种选择。程序亦是如此,也会根据不同条件执行相应的行为。例如,在"贪吃蛇"游戏中,按下不同的按键,"贪吃蛇"的运动方向随之而改变。本章主要介绍选择和循环控制结构程序设计的实现方法。

3.1 选择控制结构语句

3.1.1 if 语句

最常见的选择语句是 if 语句。if 语句用于选择是否执行一个行为,执行过程是先计算条件表达式的值,如果条件表达式为真,则执行其后的语句;否则,就直接跳过其后的语句。Java 语言提供了 3 种形式的 if 语句。

1. 单分支结构

单分支结构基本形式如下:

```
if(条件表达式){
    语句;
}
```

单分支结构非常简单,if 后面括号里的条件表达式值为真,就执行{}里的语句;否则,直接越过{}里的语句。{}里的语句是复合语句,相当于一个整体。{}里的语句可以是多条,也可以只有 1 条,当只有 1 条语句的时候,可以省略{}。

【例 3.1】 编写程序,实现通过按键控制方块运动,即当按下"w"键时,方块向上运动,如图 3.1 所示。

视频讲解

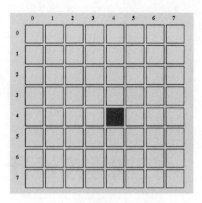

图 3.1 按键控制方块运动

为了实现按键控制方块运动，需要获得按键输入信息，然后根据判断信息执行相应的行为。Screen 类的 getKey()方法可以获得按键的输入信息，然后根据键值执行相应的操作。代码如下：

```java
public class Main {
    public static void main(String[ ] args){
        final int SIZE = 8;
        Screen screen = new Screen(SIZE);
        int row = screen.getRandom(SIZE);
        int col = screen.getRandom(SIZE);
        Cell cell = new Cell(row,col);
        screen.add(cell);
        char key;
        while(true) {
            screen.delay();
            key = screen.getKey();
            if(key == 'w') {                    // 按键为 w,向上运动
                cell.moveUp();
            }
        }
    }
}
```

运行程序，在键盘上按"w"键时，方块向上运动。有时候不小心将键盘上大写键打开了，测试程序的时候，按下"w"键，方块并没有向上运动，这个错误不易发现，所以将代码修改成按下"W"键也能向上运动，代码如下：

```java
public class Main {
    public static void main(String[ ] args){
        final int SIZE = 8;
        Screen screen = new Screen(SIZE);
        int row = screen.getRandom(SIZE);
        int col = screen.getRandom(SIZE);
        Cell cell = new Cell(row,col);
        screen.add(cell);
        char key;
        while(true) {
```

```
            screen.delay();
            key = screen.getKey();
            if(key == 'w' || key == 'W') {
                cell.moveUp();
            }
        }
    }
}
```

运行程序,当按下"W"键的时候,方块向上运动。

2. 双分支结构

除了上述单分支选择结构,Java 语言提供了 if-else 双分支语句,即在两条语句之中进行选择,格式如下:

```
if(条件表达式){
    语句 1;
}
else{
    语句 2;
}
```

if-else 语句执行过程是:如果满足条件表达式,则执行语句 1;否则执行语句 2。语句 1 和语句 2 均可以由多条语句组成。另外,需注意 else 语句不能单独使用,必须与 if 配对使用。

【例 3.2】 编写程序,实现按"W"键控制方块向上运动,按其他键控制方块向下运动。

按下"W"键向上运动,按下其他键则向下运动,这是非常明显的二选一情况,所以可以使用 if-else 语句实现要求,代码如下:

视频讲解

```
public class Main {
    public static void main(String[ ] args){
        final int SIZE = 8;
        Screen screen = new Screen(SIZE);
        int row = screen.getRandom(SIZE);
        int col = screen.getRandom(SIZE);
        Cell cell = new Cell(row,col);
        screen.add(cell);

        char key;
        while(true) {
            screen.delay();
            key = screen.getKey();
            if(key == 'w' || key == 'W') {
                cell.moveUp();
            }
            else{
                cell.moveDown();
            }
        }
    }
}
```

运行程序,当按下"W"键时,方块向上运动;按下其他键时,方块向下运动。

3. 多分支结构

在程序中,经常会遇到面临很多选择的时候,例如按下不同的按键控制方块向不同方向运动,Java 语言提供了 if-else if-else 语句,格式如下:

```
if(条件表达式 1){
    语句 1;
}
else if(条件表达式 2){
    语句 2;
}
…
else{
    语句 n;
}
```

该语句执行过程是:如果满足条件表达式 1 的时候,则执行语句 1,否则计算条件表达式 2;如果满足条件表达式 2,则执行语句 2,否则计算下一个条件表达式;直到所有条件均不满足,则执行 else 所对应的语句 n。

【例 3.3】 编写程序,实现按"W"键控制方块向上运动,按"S"键控制方块向下运动,按其他键控制方块斜向左上运动。

根据题意可知,这是多种选择的情况,可以使用 if-else if-else 语句实现功能,代码如下:

```java
public class Main {
    public static void main(String[ ] args){
        final int SIZE = 8;
        Screen screen = new Screen(SIZE);
        int row = screen.getRandom(SIZE);
        int col = screen.getRandom(SIZE);
        Cell cell = new Cell(row,col);
        screen.add(cell);
        char key;
        while(true) {
            screen.delay();
            key = screen.getKey();
            if(key == 'w' || key == 'W') {
                cell.moveUp();
            }
            else if(key == 's' || key == 'S'){
                cell.moveDown();
            }
            else{
                cell.moveLeft();
                cell.moveUp();
            }
        }
    }
}
```

运行程序,当按下"W"键时,方块向上运动;按下"S"键时,方块向下运动;按下其他按键时,方块斜向左上运动。

if 语句用于选择是否执行一个行为,if-else 语句用于在两个语句之间进行选择,if-else if-else 语句用于在多个语句之间进行选择。

3.1.2 switch 语句

对于在多个选项中选择,不仅可以用 if-else if-else 语句来完成,Java 语言还提供了一种更为方便的 switch 语句,它的格式如下:

```
switch(表达式){
    case 常量表达式 1:
        语句 1;
        break;
    case 常量表达式 2:
        语句 2;
        break;
    ...
    default:语句 n;
}
```

该语句执行过程是:先计算表达式的值,如果表达式的值与某个常量表达式的值相等,则执行其后控制的语句;如果所有的常量表达式都与表达式的值不相等,则执行 default 后的语句。

需要注意的是:

(1) case 后面必须是常量表达式,不能包含变量,并且每个常量表达式的值都不相同。

(2) default 语句可以缺省。如果省略了 default 语句,当表达式的值与所有的常量表达的值都不相等,则什么也不执行。

(3) break 语句可以终止其所在的 switch 语句的执行。switch 语句执行的原理是遇到匹配项之后,开始执行其后的语句,如果没有遇到 break 语句,会一直执行到 switch 语句结束为止。

【例 3.4】 使用 switch 语句编写程序,实现按"w"键控制方块向上运动,按"s"键控制方块向下运动,按"a"键控制方块向左运动,按键"d"控制方块向右运动。

使用 switch 语句,代码如下:

```java
public class Main {
    public static void main(String[ ] args){
        final int SIZE = 8;
        Screen screen = new Screen(SIZE);
        int row = screen.getRandom(SIZE);
        int col = screen.getRandom(SIZE);
        Cell cell = new Cell(row,col);
        screen.add(cell);
        char key;
        while(true) {
            screen.delay();
            key = screen.getKey();
            switch(key) {
                case 'w':
```

视频讲解

```
                    cell.moveUp();
                    break;
                case 's':
                    cell.moveDown();
                    break;
                case 'a':
                    cell.moveLeft();
                    break;
                case 'd':
                    cell.moveRight();
                    break;
            }
        }
    }
}
```

运行程序,"w""s""a""d"键可以分别控制方块上、下、左、右运动。

使用 swtich 语句的时候,要根据情况需要,判断是否要加上 break 语句。例如打开大写键,按键依然能控制方块上、下、左、右运动。使用 swtich 语句实现的话,代码如下:

```
public class Main {
  public static void main(String[ ] args){
    final int SIZE = 8;
    Screen screen = new Screen(SIZE);
    int row = screen.getRandom(SIZE);
    int col = screen.getRandom(SIZE);
    Cell cell = new Cell(row,col);
    screen.add(cell);

    char key;
    while(true) {
        screen.delay();
        key = screen.getKey();
        switch(key) {
            case 'w':
            case 'W':
                cell.moveUp();
                break;
            case 's':
            case 'S':
                cell.moveDown();
                break;
            case 'a':
            case 'A':
                cell.moveLeft();
                break;
            case 'd':
            case 'D':
                cell.moveRight();
                break;
        }
    }
  }
}
```

运行程序,即使打开大写键,"W""S""A""D"键依然能控制方块上、下、左、右运动。当按下按键"W"的时候,case 'w' 选项后的语句为空,程序会自动往下执行,执行 case 'W' 选项对应的程序段,所以即使大写键开启,输入的键值是"W"也能控制方块向上运动。

如果给每一个 case 选项的语句段都增加 break 语句,则需要增加不少重复的代码。对于不同条件,执行相同行为的场景下,通过 switch 语句和 break 语句的结合使用,可以写出简洁的代码。

对于 switch 语句中漏掉 break 语句这种情况时有发生,Java 14 版本采用了新的方式规避这个问题,使用简化的"case L ->"模式匹配语法,作用于不同范围并控制执行流,代码为:

```java
public class Main {
    public static void main(String[ ] args){
        final int SIZE = 8;
        Screen screen = new Screen(SIZE);
        int row = screen.getRandom(SIZE);
        int col = screen.getRandom(SIZE);
        Cell cell = new Cell(row,col);
        screen.add(cell);

        char key;
        while(true) {
            screen.delay();
            key = screen.getKey();
            switch(key) {
                case 'w','W' -> {
                    cell.moveUp();
                }
                case 's','S' -> {
                    cell.moveDown();
                }
                case 'a','A' ->{
                    cell.moveLeft();
                }
                case 'd','D' ->{
                    cell.moveRight();
                }
            }
        }
    }
}
```

将重复的 case 标签常量合并在一个 case 常量表达式中,用逗号分隔,箭头指向语句即可,执行完语句后退出 switch 语句。

虽然 switch 语句有时候比 if-else 语句简洁,逻辑关系一目了然,程序可读性好,但是使用范围较窄,比如选择条件是非常大的范围时,就不适合用 switch 语句。如表达变量 i 是 100～550 的整数时,使用 if 语句非常简单,代码如下:

```java
if( i > 100 && i < 550 )
```

而使用 switch 语句将会非常麻烦,需要设置几百个 case 选项,因此对于初学者来说,熟练掌握 if 语句即可。

3.2 循环控制结构语句

循环是指反复执行某一过程,在 3.1 节中的案例使用的 while 语句就是一种经典的循环结构。循环结构是结构化程序的基本结构之一,它与顺序结构、选择结构共同构成各种复杂的程序。换言之,所有结构化程序都是由这三种基本结构组成。Java 语言提供了 while 语句、do-while 语句和 for 语句来实现循环结构。

3.2.1 while 语句

while 语句是常见的循环语句,while 语句的格式如下:

```
while(条件表达式){
    循环体语句;
}
```

while 语句的执行过程是:当圆括号里的条件表达式为真时,执行循环体语句,然后再判断表达式里的值,如果为真,继续执行循环体语句,如此重复执行,直到表达式的值为假的时候结束循环。

while 语句和 if 语句,两者结构非常相似,两者的差别在于 if 语句只判断一次,而 while 语句会反复判断,直到条件不满足为止。

视频讲解

【例 3.5】 编写程序,使用 while 语句,实现"贪吃蛇"不断向下运动,一直运动到屏幕底部撞到墙,游戏结束,如图 3.2 所示。

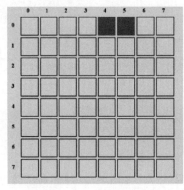 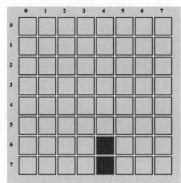

图 3.2 "贪吃蛇"向下运动

"贪吃蛇"可以不断向下运动的条件就是:"蛇头"没有达到屏幕底部。解决了判断条件,接下来就是循环体的内容,主要包括两部分:延时和向下运动。代码如下:

```java
public class Main {
    public static void main(String[ ] args){
        final int SIZE = 8;
        Screen screen = new Screen(SIZE);
        Cell head = new Cell(0,4);
```

```
        Cell body = new Cell(0,5);
        screen.add(head);
        screen.add(body);
        while(head.getRow()< SIZE - 1) {
            screen.delay();
            body.moveTo(head);
            head.moveDown();
        }
    }
}
```

运行程序,"贪吃蛇"不断向下运动,直到运动到屏幕最底部为止。

3.2.2 do-while 语句

除了 while 语句,Java 语言还提供了 do-while 语句实现循环结构,do-while 语句的格式如下:

```
do{
    循环体语句;
}while(条件表达式);
```

do-while 语句的执行过程是:先执行循环体语句,再进行判断,如果条件表达式为真,继续执行循环体语句,如此反复,直到条件表达式的值为假,结束循环。

【例 3.6】 编写程序,使用 do-while 循环语句,实现"贪吃蛇"不断向下运动,直到运动到屏幕底部。

视频讲解

使用 do-while 语句,循环执行的条件是判断"蛇头"是否运动到屏幕的底部。循环体的内容主要包括两部分,延时和向下运动,代码如下:

```
public class Main {
    public static void main(String[ ] args){
        final int SIZE = 8;
        Screen screen = new Screen(SIZE);
        Cell head = new Cell(0,4);
        Cell body = new Cell(0,5);
        screen.add(head);
        screen.add(body);

        do{
            screen.delay();
            body.moveTo(head);
            head.moveDown();
        }while(head.getRow() < SIZE - 1);
    }
}
```

运行程序,"贪吃蛇"不断向下运动,直到运动到屏幕最底部。

通常情况下,对于同一问题,既可以使用 while 语句,也可以使用 do-while 语句,两者可以互换,运行结果相同,但是两者的差别为:do-while 语句是先执行后判断,所以 do-while 语句至少执行循环体一次,而 while 语句是先判断后执行,所以可能一次也不执行循环体。因此,如果 while 语句的条件表达式一开始就为假,两个循环语句的结果是不同的。

3.2.3 for 语句

例 3.5、例 3.6 的程序都是重复执行固定次数的循环,包含如下三部分:
(1) 初始化数值。
(2) 比较数值与目标值的关系。
(3) 循环递增数值。

在 while 语句中,这三部分内容放在不同的位置,初始化数值语句放在循环外,比较数值语句放在 while 语句中括号里面,而递增数值语句放在循环体语句内。当循环体里语句较多时,有时候会遗忘递增数值,从而导致循环无限执行。

Java 语言还提供了 for 循环语句,将初始化、判断、更新数值三部分内容组合在一起,形成更加紧凑的形式。for 循环语句基本格式如下:

```
for (表达式 1; 表达式 2; 表达式 3){
    循环语句;
}
```

表达式 1 通常是为控制循环的变量赋初值;表达式 2 是判断循环是否执行的条件表达式;表达式 3 一般用来更新控制循环的变量,其后无分号。

for 语句执行过程如下:
(1) 执行"表达式 1",为控制循环的变量赋初值。
(2) 判断"表达式 2",如果它的值为真,则执行循环体内的语句,否则转到第(5)步。
(3) 执行"表达式 3",更新变量。
(4) 重复执行步骤(2)和(3)。
(5) 结束循环。

for 语句是 while 语句的一种变体,比 while 语句使用起来更加灵活。

【例 3.7】 编写程序,使用 for 循环语句,实现"贪吃蛇"不断向下运动,直到运动到屏幕底部。

使用 for 语句实现功能,代码如下:

视频讲解

```java
public class Main {
    public static void main(String[ ] args){
        final int SIZE = 8;
        Screen screen = new Screen(SIZE);
        Cell head = new Cell(0,4);
        Cell body = new Cell(0,5);
        screen.add(head);
        screen.add(body);

        for(int row = head.getRow(); row < SIZE - 1; row ++) {
            screen.delay();
            body.moveTo(head);
            head.moveDown();
        }
    }
}
```

运行程序,"贪吃蛇"不断向下运动,直到运动到屏幕最底部为止。

与 while 循环语句相比，for 循环语句结构更为紧凑，不容易遗忘更新变量，尤其是多重循环的时候，优势更加明显，因此 for 循环语句的使用非常广泛。

3.2.4 三种循环的比较

通过例 3.5～例 3.7 可知，一般情况下，对于同一问题，while、do-while、for 循环语句都可以处理，三者可以相互替代。

for 与 while 循环语句是"当型"循环，即先判断条件后执行循环体。do-while 循环语句是"直到型"循环，即先执行循环体，后判断条件。因此，for 与 while 循环语句可能一次也不执行循环体的内容，而 do-while 循环语句至少执行一次循环体的内容。

如何选择哪一种循环来解决问题？如果循环次数明确的时候，使用 for 循环更为方便，而循环次数不确定的时候，while 循环语句更容易理解。如果第一次循环肯定会执行，可以使用 do-while 循环语句。

for 循环语句比 while 和 do-while 循环语句功能更强大，使用更灵活。for 循环语句中的三个表达式可以部分省略或全部省略，但是两个分号不能省略，而且语句的位置可以灵活多变。

例如：

```
for( int count = 0; count < 6; count ++ ){
}
```

可以写成如下格式：

```
for( int count = 0; count < 6;){
     count ++ ;
}
```

或者

```
int count = 0;
for( ; count < 6; count ++ ){
}
```

还有其他多种写法，不一一列举，所以 for 语句非常灵活。

3.2.5 嵌套循环语句

一个循环体内包含着另一个循环结构，称为嵌套循环。嵌套循环里还可以继续嵌套循环，循环层次可以不断叠加。三种循环语句均可以相互嵌套。

【例 3.8】 编写程序，实现"打砖块"游戏中"砖墙"功能，将屏幕上面 3 行都填满方块，如图 3.3 所示。

要将屏幕的上 3 行都填满方块，可以使用双重循环实现，内层循环的作用是将屏幕中的某一行都填满方块，外层循环的作用是控制行，实现从第 0 行到 2 行逐行填满方块，代码

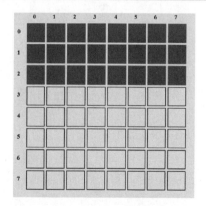

图 3.3　屏幕上 3 行填满方块

如下:

```java
public class Main {
    public static void main(String[ ] args){
        final int SIZE = 8;
        Screen screen = new Screen(SIZE);

        for(int row = 0; row < 3; row ++ ) {
            for(int col = 0; col < SIZE; col ++ ) {
                Cell cell = new Cell(row,col);
                screen.add(cell);
            }
        }
    }
}
```

运行程序,屏幕上3行都填满了方块。在阅读嵌套循环的时候,就像看钟表一样,内层循环像秒钟转了一圈,外层循环像分钟转一格。

对于例3.8也可以使用两个while循环嵌套实现,代码如下:

```java
public class Main {
    public static void main(String[ ] args){
        final int SIZE = 8;
        Screen screen = new Screen(SIZE);

        int row = 0;
        while(row < 3) {
            int col = 0;
            while(col < SIZE) {
                Cell cell = new Cell(row,col);
                screen.add(cell);
                col ++ ;
            }
            row ++ ;
        }
    }
}
```

对比while和for循环语句,可以看出,for循环语句的结构更为紧凑,而且不容易遗忘变量的更新。

3.2.6 break语句和continue语句

1. break语句

在循环中,有时候需要提前结束循环,就像游戏进行到一半时,想提前结束游戏,Java语言提供break语句,可以提前结束循环。

break语句的一般格式如下:

```
break;
```

【例 3.9】 编写程序,实现按"W"键控制方块向上运动,按"S"键控制方块向下运动,按"K"键结束游戏。

相比例 3.3 和例 3.4 用按键控制方块运动的任务,该任务多了新功能,即按"K"键结束游戏。需要提前结束循环,则可以使用 break 语句,代码如下:

```java
public class Main {
    public static void main(String[ ] args){
        final int SIZE = 8;
        Screen screen = new Screen(SIZE);
        int row = screen.getRandom(SIZE);
        int col = screen.getRandom(SIZE);
        Cell cell = new Cell(row,col);
        screen.add(cell);
        char key;
        while(true) {
            screen.delay();
            key = screen.getKey();
            if(key == 'w' || key == 'W') {
                cell.moveUp();
            }
            if(key == 's' || key == 'S'){
                cell.moveDown();
            }
            if(key == 'k' || key == 'K'){
                break;
            }
        }
    }
}
```

运行程序,按键"W"控制方块向上运动,按键"S"控制方块向下运动,按键"K"控制游戏结束。

2. continue 语句

continue 语句也可以提前结束循环,不过它只结束本次循环,即跳过本次循环剩下的部分,进入下一次循环。

continue 语句的一般格式如下:

```
continue;
```

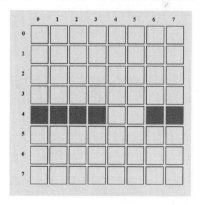

图 3.4 赛车障碍物

【例 3.10】 "赛车类"游戏中会出现一排障碍物,中间只留下狭窄的空隙供赛车通过,如图 3.4 所示。

对于障碍物来说,除了"空隙"外,该行其他位置都应添加方块。因此,可以使用循环语句逐一添加方块,"空隙"的位置使用 continue 语句,跳过本次循环,代码如下:

```java
public class Main {
    public static void main(String[ ] args){
        final int SIZE = 8;
        Screen screen = new Screen(SIZE);
```

```
            int row = screen.getRandom(SIZE);
            int gapCol = screen.getRandom(SIZE - 1);
            for(int col = 0; col < SIZE; col ++ ) {
                if(col == gapCol || col == gapCol + 1) {
                    continue;
                }
                Cell cell = new Cell(row,col);
                screen.add(cell);
            }
        }
    }
```

运行程序,显示如图 3.4 所示的界面。对于例 3.10,也可以不使用 continue 语句实现,代码如下:

```
public class Main {
    public static void main(String[ ] args){
        final int SIZE = 8;
        Screen screen = new Screen(SIZE);
        int row = screen.getRandom(SIZE);
        int gapCol = screen.getRandom(SIZE - 1);

        for(int col = 0; col < SIZE ; col ++ ) {
            if(col! = gapCol && col ! = gapCol + 1) {
                Cell cell = new Cell(row,col);
                screen.add(cell);
            }
        }
    }
}
```

通过案例可知,程序设计非常灵活,同一个问题可以有多种解决办法。continue 语句和 break 语句的区别是:break 语句是终止整个循环过程,而 continue 语句只结束本次循环,而且 continue 语句只能用在循环语句之中。

视频讲解

3.3 综合案例:按键控制"贪吃蛇"运动

编写程序,实现按键控制"贪吃蛇"运动,按键"w"控制"贪吃蛇"向上运动,按键"s"控制"贪吃蛇"向下运动,按键"a"控制"贪吃蛇"向左运动,按键"d"控制"贪吃蛇"向右运动,如图 3.5 所示。

"贪吃蛇"的运动规律,主要由两部分组成:"蛇头"和"蛇身","蛇头"运动规律与运动方向有关,"蛇身"运动规律与运动方向无关,无论什么运动方向,"蛇身"的每一部分都会移动到与之相邻方块的位置。

代码如下:

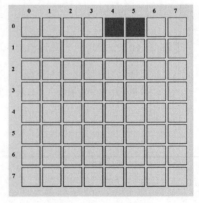

图 3.5 "贪吃蛇"运动

```java
public class Main {
    public static void main(String[ ] args){
        final int SIZE = 8;
        Screen screen = new Screen(SIZE);
        Cell head = new Cell(0,4);
        Cell body = new Cell(0,5);
        screen.add(head);
        screen.add(body);

        char key;
        while(true) {
            screen.delay();
            key = screen.getKey();
            if(key == 'w') {
                body.moveTo(head);
                head.moveUp();
            }
            if(key == 's') {
                body.moveTo(head);
                head.moveDown();
            }
            if(key == 'a') {
                body.moveTo(head);
                head.moveLeft();
            }
            if(key == 'd') {
                body.moveTo(head);
                head.moveRight();
            }
        }
    }
}
```

提示：计算机相比于人最擅长的就是做重复的事，很多软件设计的目的就是解决重复劳动的问题。

习题

3.1 下列程序段，关于循环执行的次数，说法正确的是_____。

```
int k = 5 ;
while(k > 1){
    k -- ;
}
```

  A. 循环执行了 5 次   B. 循环执行了 0 次

  C. 循环执行了 4 次   D. 循环执行了 3 次

3.2 执行以下程序，sum 的值为_____。

```
int sum = 0;
int i ;
for( i = 0; i <= 5; i++){
    sum + = i;
}
```

  A. 10   B. 15   C. 21   D. 28

3.3 编写程序,在屏幕上显示杨辉三角图形,如图3.6所示。

第0行添加1个方块;

第1行添加2个方块;

…

第N行添加N+1个方块。

3.4 编写程序,在屏幕上显示倒杨辉三角图形,如图3.7所示。

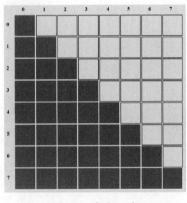
图3.6 杨辉三角

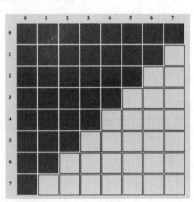
图3.7 倒杨辉三角

3.5 编写程序,按键"W""S""A""D"控制"Z"形"俄罗斯方块"(如图3.8所示)上、下、左、右运动。

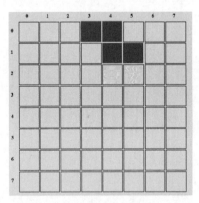
图3.8 "Z"形"俄罗斯方块"示意图

第4章 数　　组

在"贪吃蛇"游戏中,当"贪吃蛇"的长度只有两个方块的时候,可以使用两个 Cell 类型的变量去存储方块的信息。此时数据量较小,这种方式尚能应付。但是当数据量较大时,比如"贪吃蛇"的长度为几十个方块时,就不能采用这种方法了。处理大量相关数据的时候,就需要更好的方式进行存储和处理。在生活中如果数据较多,人们可以采用表格来记录数据,这样处理数据会更便捷。Java 语言也提供了一种类似表格的结构来存储数据,被称为数组,数组能高效、便捷地处理数据。

4.1 一维数组

数组是同类型有序数据的集合,其中的每个元素都是相同的数据类型。例如,数组有 10 个元素,这 10 个元素都必须是同一数据类型。数组存储在一段连续的内存空间上,可以通过数组名称加索引(也被称为下标或者偏移量)访问数组中的元素,如图 4.1 所示。

图 4.1　数组存储数据示意图

4.1.1 一维数组的定义

在 Java 语言中,数组是引用数据类型,在使用之前需要声明和初始化。声明数组就是告诉编译器数组名和数组元素类型。声明一维数组的格式有如下两种:

```
类型说明符 [ ] 数组名;
类型说明符 数组名[ ];
```

建议使用第一种方式,更加符合 Java 语言习惯,第二种方式是为了兼顾 C 语言数组使用习惯。

数据类型可以是基础类型,也可以是引用类型,例如:

```
int [ ] name;
Cell [ ] cells;
```

数组 name 中,每个元素都可以存储 int 类型的值。而数组 cells 中,每个元素都可以存储 Cell 类型的值。如果数组元素为引用类型,则该数组为对象数组,数组 cells 就是对象数组。

声明数组的时候,需要指明元素的类型,并且给数组命名,数组名的命名规则与变量名的命名规则一致。声明数组后,还必须为它分配内存空间,为一维数组分配内存空间的格式如下:

```
数组名字 = new 数组元素的类型[数组元素的个数];
```

例如:

```
name = new int [5];
```

也可以在声明数组的时候就给它分配空间,例如:

```
int [ ] name = new int[5];
Cell [ ] cells = new Cell[5];
```

需要注意的是,一旦数组初始化之后,数组的大小就不能改变。

4.1.2 一维数组的初始化

数组分配完内存空间之后,需要对其初始化,也就是为每个元素赋初值。数组的初始化通常有两种方式:动态初始化和静态初始化。

1. 动态初始化

动态初始化是指在初始化数组时由程序设计者指定数组的长度,然后由系统为数组元素分配初始值。系统根据指定的长度分配内存空间供数组使用,并且根据数组元素的数据类型设置默认值。整型数据默认初始值为"0",布尔型数据默认初始值为"false",引用型数据默认初始值为"null"。

例如:

```
int [ ] cols = new int [5];
```

表示数组 cols 有 5 个 int 类型的元素,并且每个元素的初始值都是"0"。

```
Cell [ ] cells = new Cell[4];
```

表示数组 cells 有 4 个 Cell 类型的元素,并且每个元素的初始值都是"null"。

2. 静态初始化

静态初始化指的是初始化数组时由程序设计者为每个元素赋初值。静态初始化包括两种方式:完整方式和简化方式。

(1) 完整方式的语法格式如下:

```
数据类型[ ] 数组名称 = new 数据类型[ ]{值,值,…};
```

(2) 简化方式的语法格式如下:

```
数据类型[ ] 数组名称 = {值,值,…};
```

例如:

```
int [ ] cols = new int[ ] {1,2,3,4,5};
```

其简化方式为:

```
int [ ] cols = {1,2,3,4,5};
```

表示数组 cols 有 5 个 int 型元素,初始值分别为 1,2,3,4,5。

4.1.3 一维数组的使用

1. 数组元素的引用

需要使用数组中的元素,可以通过数组名称加下标(也被称为索引)的方式进行访问,引用数组元素的格式如下:

```
数组名[下标];
```

Java 语言中数组下标的值必须是整数,并且从 0 开始计数,例如:

```
int [ ] cols = {6,5,4,3,2 };
```

cols[0]表示第 1 个元素,值为 6;cols[1]表示第 2 个元素,值为 5,以此类推,如图 4.2 所示。

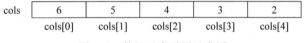

图 4.2 数组元素引用示意图

在生活中,常常遇到通过索引值获得相应内容的情形。例如,老师在上课的时候,常常会讲到翻看数学书的第 40 页,数学书就相当于数组名,不同的数组名意味着不同的书,而第 40 页中的 40 就是索引值。从数组中引用元素的过程与这类似,通过数组名找到对应的数组,然后通过索引找到对应位置的值。例如:

```
cols[0] = col[1] + cols[2];
```

表示将 cols[1]的值 5 与 cols[2]的值 4 相加,得到的结果存储到 cols[0]中,所以 col[0]的值变为 9。

使用数组时,要防止数组下标越界。例如,数组 cols 有 5 个元素,使用该数组的时候,要确保下标范围在 0～4。

【例 4.1】 编写程序,通过数组将"T"形"俄罗斯方块"显示在屏幕上,如图 4.3 所示。

使用数组分别存储四个方块的信息,然后将每个方块显示在屏幕上,就能显示出"T"形"俄罗

视频讲解

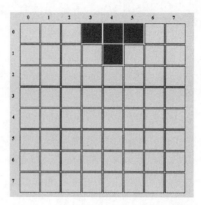

图 4.3 "T"形"俄罗斯方块"示意图

斯方块"的形状,代码如下:

```java
public class Main {
    public static void main(String[ ] args){
        final int SIZE = 8;
        Screen screen = new Screen(SIZE);

        Cell[ ]cells = new Cell[4];
        cells[0] = new Cell(0,3);
        cells[1] = new Cell(0,4);
        cells[2] = new Cell(0,5);
        cells[3] = new Cell(1,4);
        screen.add(cells);
    }
}
```

运行程序,屏幕上显示出"T"形"俄罗斯方块"的形状。

2. 数组的遍历

数组的使用中,经常需要遍历数组,即访问数组中的每个元素。例如,"俄罗斯方块"游戏中的方块不断向下运动,就需要遍历数组中的每一个元素,让每一个方块都向下运动。数组的遍历可以通过循环语句来实现,设置一个变量存储数组的下标,变量的值从 0 开始逐步递增至数组的大小减 1 就能遍历数组。数组的大小可以通过成员变量 length 获得,例如 cells.length 就是获得数组 cells 的长度。遍历数组 cells 实现"俄罗斯方块"向下运动的完整代码如下:

```java
public class Main {
    public static void main(String[ ] args){
        final int SIZE = 8;
        Screen screen = new Screen(SIZE);

        Cell [ ] cells = new Cell[4];
        cells[0] = new Cell(0,3);
        cells[1] = new Cell(0,4);
        cells[2] = new Cell(0,5);
        cells[3] = new Cell(1,4);
        screen.add(cells);
        screen.delay();
```

```
        for(int i = 0; i < cells.length; i ++ ){
            cells[i].moveDown();
        }
    }
}
```

运行程序,"俄罗斯方块"向下运动一格。

3. 增强型 for 循环

Java 语言还提供了一种新的循环类型,它能在不使用下标的情况下遍历数组,被称为 for-Each 循环或者加强型循环。语法格式如下:

```
for(元素数据类型 元素:数组名){
}
```

使用增强型 for 循环,遍历数组 cells 实现"俄罗斯方块"向下运动的代码如下:

```
public class Main {
    public static void main(String[ ] args){
        final int SIZE = 8;
        Screen screen = new Screen(SIZE);

        Cell [ ] cells = new Cell[4];
        cells[0] = new Cell(0,3);
        cells[1] = new Cell(0,4);
        cells[2] = new Cell(0,5);
        cells[3] = new Cell(1,4);
        screen.add(cells);
        screen.delay();
        for(Cell cell:cells){
            cell.moveDown();
        }
    }
}
```

运行程序,"俄罗斯方块"向下运动一格。

4.2 二维数组

"俄罗斯方块"游戏中,共有 7 种不同形状的方块。如果每一种形状都使用一个一维数组保存方块的位置信息,则总共需要 7 个一维数组保存 7 种形状。虽然该方案可以解决问题,但是代码会变得非常烦琐。例如,随机产生某种形状的方块,则需要使用多条选择语句,根据不同的条件产生对应形状的方块。如果使用二维数组就能简化代码,二维数组与一维数组相似,一维数组是多个相同类型元素的集合,而二维数组就是多个类型和大小都相同的一维数组的集合,因此,二维数组也被称为数组的数组。

4.2.1　二维数组的定义

二维数组声明的一般格式如下：

```
类型说明符[ ] [ ] 数组名;
```

声明数组后，还必须为它分配内存空间，为二维数组分配内存空间的格式如下：

```
数组名字 = new 数组元素的类型[行数][列数];
```

例如：

```
int [ ][ ] value = new int [3][4];
```

表示数组 value 是一个二维数组，由 3 个一维数组组成，分别是 value[0]、value[1]、value[2]，每个一维数组有 4 个元素，数组共有 3×4=12 个数组元素，如图 4.4 所示。

value[0]	value[0][0]	value[0][1]	value[0][2]	value[0][3]
value[1]	value[1][0]	value[1][1]	value[1][2]	value[1][3]
value[2]	value[2][0]	value[2][1]	value[2][2]	value[2][3]

图 4.4　二维数组示意图

与一维数组一样，二维数组在内存中也是按顺序存放的，先存放第 0 行的所有元素，接着存放第 1 行所有元素，以此类推。

4.2.2　二维数组的初始化

二维数组的初始化也有两种方法：动态初始化和静态初始化。
(1) 动态初始化的格式如下：

```
数据类型 [ ][ ] 数组名 = new 数据类型[行数][列数]
```

例如：

```
Cell[ ][ ] cells = new Cell[3][4];
```

表示数组 cells 是一个二维数组，共有 3 行 4 列，数组中每一个元素的数据类型都是 Cell 类型，初始值为默认的"null"。

(2) 静态初始化的格式如下：

```
数据类型 [ ][ ] 数组名 = {{元素 1,元素 2…},{元素 1,元素 2…},{元素 1,元素 2…}…};
```

例如：

```
int [ ][ ] value = {{5,4,3,2},{1,2,4,5},{3,2,1,4}};
```

表示数组 value 是一个二维数组，共有 3 行 4 列，数组中每个元素的数据类型都是 int 类型。第 1 行的初始值分别为 5、4、3、2，第 2 行的初始值分别为 1、2、4、5，第 3 行的初始值分别为 3、2、1、4。

4.2.3 二维数组的引用

如果要访问二维数组中某个元素,可以通过数组名加行下标和列下标实现,格式如下:

数组名[行下标][列下标]

与一维数组一样,行、列下标都是从 0 开始。例如:

int[][]　　value = {{0,1,2,3},{4,5,6,7},{8,9,10,11}};

value[0][3]的值为 3,value[1][2]的值为 6,而 value[3][2]就是错误引用,行下标只能为 0、1、2,当行下标值为 3 就越界了。

【例 4.2】 编写程序,随机显示一种"俄罗斯方块"的形状,"俄罗斯方块"的 7 种形状如图 4.5 所示。

视频讲解

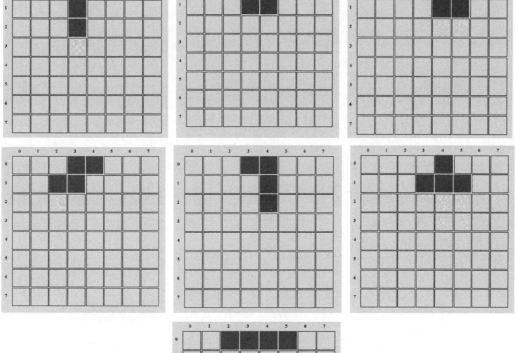

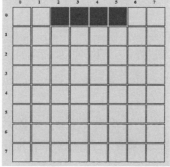

图 4.5 "俄罗斯方块"的 7 种形状示意图

使用二维数组保存7种形状的位置信息,然后产生一个0~6的随机整数,根据这个数值,从二维数组中获得相应的位置信息生成方块对象,就能显示出对应的形状,代码如下:

```java
public class Main {
    public static void main(String[ ] args){
        final int SIZE = 8;
        final int NUM = 7;
        Screen screen = new Screen(SIZE);
        int [ ][ ]rows = {{0,0,1,2},
                         {0,0,1,1},
                         {0,0,1,1},
                         {0,0,1,1},
                         {0,0,1,2},
                         {0,1,1,1},
                         {0,0,0,0}};
        int [ ][ ]cols = {{3,4,3,3},
                         {3,4,3,4},
                         {3,4,4,5},
                         {3,4,3,2},
                         {3,4,4,4},
                         {4,3,4,5},
                         {2,3,4,5}};
        Cell[ ] cells = new Cell[4];

        int index = screen.getRandom(NUM);
        for(int i = 0; i < 4; i++) {
            cells[i] = new Cell(rows[index][i],cols[index][i]);
        }
        screen.add(cells);
    }
}
```

运行程序,会随机出现一种"俄罗斯方块"的形状。

视频讲解

4.3 综合案例:"贪吃蛇"游戏

"贪吃蛇"是一款非常经典的休闲益智类游戏,玩法非常简单,通过上、下、左、右键控制"蛇"的运动方向,使"蛇"可以吃到"食物"。吃到"食物"之后,"蛇"会变得越来越长。如果撞上自己的身体或者墙壁,游戏就结束。

本例需要实现的功能是按键控制"蛇"上、下、左、右运动,如图 4.6 所示,上面四个方块构成的是"贪吃蛇"。

1. 初始化"贪吃蛇"

首先,需要找到合适的数据类型存储游戏元素。"贪吃蛇"由一组方块组成,可以使用一维数组来存储"贪吃蛇"的位置信息。保存"贪吃蛇"数据的数组如下:

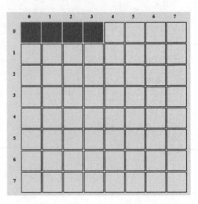

图 4.6 "贪吃蛇游戏"示意图

```
Cell [ ] snake = new Cell[4];
```

2. 显示"贪吃蛇"

初始化数据之后,根据数据信息,可以将"贪吃蛇"显示在屏幕上,代码如下:

```
public class Main {
    public static void main(String[ ] args){
        final int SIZE = 8;
        Screen screen = new Screen(SIZE);

        Cell[ ] cells = new Cell[4];                    // 初始化"蛇"的信息
        int len = cells.length;
        for(int i = 0; i < len; i++ ) {
            cells[i] = new Cell(0,i);
            screen.add(cells[i]);
        }
    }
}
```

运行程序,"贪吃蛇"显示在屏幕上。

3. 按键控制"贪吃蛇"运动方向

接下来实现用"w""s""a""d"键分别控制"贪吃蛇"向上、向下、向左、向右运动。第 3 章的综合案例中实现的"贪吃蛇"游戏,与经典版的"贪吃蛇"游戏有一点区别,那就是经典版中的"贪吃蛇"会一直沿着某个方向运动,按键按下改变其运动方向。解决这个问题的方法是定义一个变量,记录"贪吃蛇"运动的方向,值"1""2""3""4"分别代表方向上、下、左、右。如果有方向键按下,"贪吃蛇"的运动方向根据按键发生变化,否则就按原方向继续运动。"贪吃蛇"运动规则同样分为"蛇身"和"蛇头","蛇身"运动到前一个方块位置,"蛇头"根据按键指示方向运动,代码如下:

```
public class Main {
    public static void main(String[ ] args){
        final int SIZE = 8;
        Screen screen = new Screen(SIZE);
        Cell[ ] cells = new Cell[4];
        int len = cells.length;

        for(int i = 0; i < len; i++ ) {
            cells[i] = new Cell(0,i);
            screen.add(cells[i]);
        }

        int dir = 2;                    // 初始方向向下
        char key;
        while(true) {
            screen.delay();
            for(int i = len - 1; i > 0; i-- ) {
                cells[i].moveTo(cells[i-1]);
            }
```

```
            if(dir == 1) {
                cells[0].moveUp();
            }
            if(dir == 2) {
                cells[0].moveDown();
            }
            if(dir == 3) {
                cells[0].moveLeft();
            }
            if(dir == 4) {
                cells[0].moveRight();
            }
            key = screen.getKey();
            if(key == 'w') {
                dir = 1;
            }
            if(key == 's') {
                dir = 2;
            }
            if(key == 'a') {
                dir = 3;
            }
            if(key == 'd') {
                dir = 4;
            }
        }
    }
}
```

运行程序,"贪吃蛇"会一直沿着某个方向运动,除非通过按键改变其运动方向。

实现了简易版的"贪吃蛇"小游戏,还缺少"贪吃蛇"吃到"食物"身体变长的功能,这个功能会在第 7 章实现。

前面 4 章是 Java 编程的基础,与 C 语言基础有着许多相似的地方,接下来的章节主要讲解面向对象相关的知识。

习题

4.1 若定义了"int [] a={1,2,3};",数组 a 共有_____个元素。
   A. 3   B. 11   C. 9   D. 10

4.2 若定义了"int [][] a={{1,2,3},{4,5,6}};",元素 a[1][2]的值为_____。
   A. 2   B. 4   C. 5   D. 6

4.3 编写程序,实现按键控制"赛车"方块上、下、左、右移动,如图 4.7 所示。

4.4 编写程序,实现 8 个方块随机出现在屏幕上,如图 4.8 所示。

4.5 编写程序,实现按键"K"控制"T"形"俄罗斯方块"旋转,如图 4.9 所示。

4.6 编写程序,实现"俄罗斯方块"游戏中消行功能。即当一行满行之后,消掉该行,且该行上面的方块都往下移动一行,如图 4.10 所示。

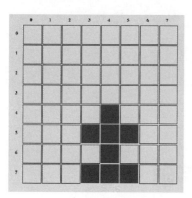
图 4.7 "赛车"示意图

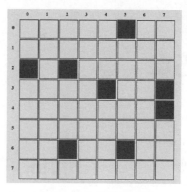
图 4.8 8个方块示意图

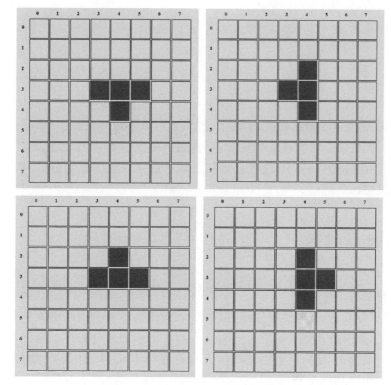
图 4.9 "俄罗斯方块"旋转示意图

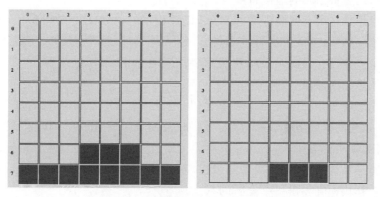
图 4.10 "俄罗斯方块"消行示意图

第5章

面向对象基础

前面 4 章是 Java 语言的基础内容,掌握了之后,就可以实现"贪吃蛇""俄罗斯方块"等经典小游戏的基础功能。但是随着功能越来越复杂,面临的挑战也越来越大,因此需要学习面向对象相关的内容来应对。在前面的章节中,读者已经接触到了面向对象的思想,接下来的章节将进一步介绍面向对象的思想。

5.1 面向对象概述

面向对象思想是在结构化设计方法出现很多问题的情况下应运而生的。随着问题规模越来越大、应用系统越来越复杂、需求变更越来越频繁,结构化设计方法面临着维护和扩展的困难,甚至一个微小的变动,都会波及整个系统。这驱使人们寻求一种新的程序设计方法,以适应现代社会对软件开发的更高要求,面向对象由此产生。面向对象是一种符合人类思维习惯的编程思想,现实生活中人们习惯将物体进行分类。例如,见到一只狗的时候,第一时间会将其进行归类,柯基、泰迪或者其他,然后根据其类型进行反馈。

面向对象的思想是将所有问题都看作对象,设计程序的时候无论要实现什么功能,首先就是要找到合适的对象。例如,生活中执行洗衣的任务,无论是通过保洁人员,还是通过洗衣机清洗,都是通过具体对象去实现。面向对象将功能封装进对象之中,然后让对象去实现具体的细节,这非常符合人类自然思考的习惯。

面向对象的思想使复杂的事情变得简单化,程序设计者无须关注实现的细节,只需要通过对象去完成即可,从以前的执行者变成指挥者。例如,前面章节实现的"贪吃蛇"游戏中方块移动的功能,程序设计者无须了解方块上、下、左、右运动是如何实现,只需要会使用相应的对象即可。面向对象更利于对复杂系统进行分析、设计与编程,能有效提高编程的效率,通过封装技术及消息机制可以像搭积木一样快速开发出一个全新的系统。

5.2 类和对象

5.2.1 对象的创建与使用

类和对象是面向对象思想中两个最基本的概念,类是对某一类事物的抽象描述,是现实世界或思维世界中的实体在计算机中的反映。而对象用于表示现实中该类事物的实例个体。简而言之,类是对象的模板,对象是类的实例,一个类可以对应多个对象。

在面向对象的编程思想中,一切皆为对象。在程序中创建对象,必须先定义一个类,然后通过类生成具体的对象。Java 语言使用 new 关键字来创建对象,格式如下:

```
类名 对象名称 = new 类名();
```

例如,创建 Cell 类的对象代码如下:

```
Cell cell = new Cell(2,4);
```

上述代码中,"Cell cell"声明了一个 Cell 类型的变量 cell,"new Cell(2,4)"用于创建 Cell 类的对象。

成功创建对象之后,可以访问对象的成员,格式如下:

```
对象引用.对象成员
```

例如:

```
cell.moveUp();
```

上述代码访问了变量 cell 的成员方法"moveUp()",作用是使方块向上运动一格。

为了对类和对象有更直观的感受,"模拟电子屏"项目中新增加了 HuskyDog 类和 TeddyDog 类,通过这两个类可以生成对象"哈士奇"和"泰迪",并且有对应的成员方法,用于设置它们在屏幕上的位置。

【例 5.1】 编写程序,将"泰迪""哈士奇"显示在屏幕上,如图 5.1 所示。

屏幕上有两种不同类型的狗,分别是"哈士奇"和"泰迪",项目提供了 HuskyDog 类和 TeddyDog 类,通过这两个类可以产生具体的对象,然后将其显示在屏幕上,代码如下:

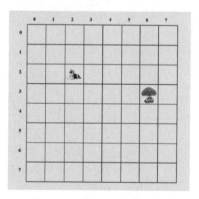

图 5.1 在屏幕上显示两种狗

```
public class Main {
    public static void main(String[ ] args){
        final int SIZE = 8;
```

```
        Screen screen = new Screen(SIZE);

        HuskyDog husky = new HuskyDog();          //创建哈士奇对象
        husky.setPoint(2,2);                       //设置在屏幕上的位置
        screen.add(husky);

        TeddyDog teddy = new TeddyDog();
        teddy.setPoint(3,6);
        screen.add(teddy);
    }
}
```

通过上述例子可以看到,类就是创建对象的模板,不同的类会产生不同的对象。如果想在屏幕上显示"哈士奇",就通过 HuskyDog 类产生具体的对象;如果想在屏幕上显示"泰迪",就通过 TeddyDog 类产生具体的对象。

5.2.2 类的定义

在掌握了如何通过已经设计好的类创建对象之后,接下来就要学习如何设计类。类是对象的抽象概念用来描述一组对象的共同特征和行为。如果想创建对象,首先必须定义类。定义一个新的类,就创建了一种新的数据类型。在学习设计类之前,先看看已经设计好的类有哪些组成部分,以 HuskyDog 类为例,代码如下:

```
public class HuskyDog{
    private int row;
    private int col;
    public void setPoint(int r, int c){
        row = r;
        col = c;
    }
    public void moveUp(){
        row -- ;
    }
    public void moveDown(){
        row ++ ;
    }
    public void moveLeft(){
        col -- ;
    }
    public void moveRight(){
        col ++ ;
    }
    public int getRow() {
        return row;
    }
    public int getCol() {
        return col;
    }
}
```

程序定义了一个 HuskyDog 类表示"哈士奇"狗,在该类中定义了两个变量 row、col 分别表示

"哈士奇"在屏幕上的行、列信息。另外，定义了 7 个方法，其中方法 setPoint()用于设置狗在屏幕上的位置，方法 moveUp()、moveDown()、moveLeft()、moveRight()用于控制狗上、下、左、右运动，方法 getRow()和 getCol()用于获得位置信息。通过上述例子可知，类主要包含属性和方法两部分。

（1）属性也称为成员变量，用来描述类的静态特征，比如位置信息、狗的品种、大小等状态信息。

（2）方法也称为成员方法，用来描述类的动态特征，包括功能或者行为，比如狗的上、下、左、右运动。

熟悉了类的基本组成部分之后，接下来对类的定义格式、类的成员变量和成员方法进行详细讲解。

1. 类定义格式

Java 语言中，类是通过 class 关键字来定义，一般格式如下：

```
[修饰符] class 类名[extends 父类名][implements 接口名]{
    // 成员变量声明
    // 成员方法声明
}
```

1）类的修饰符

类的修饰符用于说明类的特殊性质，主要分为两类：访问修饰符和非访问修饰符。访问修饰符能够有效进行访问控制，实现程序"高内聚，低耦合"的目标。访问控制一方面防止用户接触到不该接触的内容，另一方面将变与不变的部分分离，允许库设计者改变类的内部工作机制，而不必担心会对使用者产生影响。

class 前面的访问修饰符可以是 public 关键字或者默认。若修饰符是 public，则该类是公共类，可以被任何包中的类使用。若为默认，也就是没有显性设置访问控制符，则该类只能被同一包其他类使用，因此这一访问特性又称为包访问性。

对于非访问修饰符，如抽象类修饰符 abstract 和最终类修饰符 final 会在后面的章节详细介绍。

2）类名

Java 类命名符合标识符的命名规则，并且约定类名首字母要大写。

3）extends 和 implements

关键字 extends 和 implements 是可选项，extends 用于说明所定义的类继承于哪个父类。implements 用于说明当前类实现了哪些接口。extends 和 implements 的内容将在下一章详细讲解。

2. 声明成员变量

类声明结束后是一对大括号，大括号里的内容称为类体，主要包括类的成员变量和成员方法。成员变量的声明格式如下：

```
[修饰符] 数据类型 变量名 [ = 值];
```

成员变量的修饰符也分为两类：访问修饰符和非访问修饰符。修饰符会在后文中详细介绍。

数据类型可以为 Java 中的任意类型，既可以为基本数据类型，也可以为引用类型。变量名的命名必须符合标识符的命名规则。在声明变量的时候可以赋初值，也可以不赋初值。

3. 声明成员方法

成员方法是类的行为特征，类似于 C 语言中的函数，方法的声明如下：

```
[修饰符] 返回值类型 方法名([参数列表]){
}
```

1) 方法的修饰符

方法的修饰符也分为两类：访问修饰符和非访问修饰符。修饰符是可选项，一个方法如果缺省访问修饰符，则可以被同一个类的方法访问或者同一个包中的类访问。方法的修饰符较多，有静态修饰符 static、最终修饰符 final 等，这些修饰符会在后面的内容中逐一介绍。

2) 返回值类型

返回值类型为方法返回值的数据类型，可以是任何数据类型，若一个方法没有返回值，则使用关键字 void。

3) 方法名

方法名的命名规则符合标识符命名规则。

4) 参数列表

参数列表是调用方法时传递的参数信息，方法可以没有参数，也可以有 1 个或多个参数。如果有多个参数，参数声明中间用逗号分隔，而且每个参数都必须包含数据类型和参数名。

5) 返回值

当方法需要返回数据的时候，使用 return 语句。当方法不需要返回数据的时候，可以省略 return 语句。

熟悉了类的定义方式之后，接下来通过具体案例来进一步掌握类的定义。

视频讲解

【例 5.2】 为 HuskyDog 类增加一个移动到屏幕某个位置的方法 moveTo(int r,int c)。

新增成员方法 moveTo()，作用是将"哈士奇"移动到屏幕某个位置，也就是将其成员变量 row 和 col 的值设置为参数传递的值，代码如下：

```java
public class HuskyDog{
    private int row;
    private int col;
    public void setPoint(int r, int c){
        row = r;
        col = c;
    }
    public void moveUp(){
        row -- ;
    }
    public void moveDown(){
        row ++ ;
    }
```

```java
    public void moveLeft(){
        col -- ;
    }
    public void moveRight(){
        col ++ ;
    }
    public void moveTo(int r, int c){
        row = r;
        col = c;
    }
    public int getRow() {
        return row;
    }
    public int getCol() {
        return col;
    }
}
```

给 HuskyDog 类增加了 moveTo()方法,可以通过该方法控制"哈士奇"的移动,测试代码如下:

```java
public class Main {
    public static void main(String[ ] args){
        final int SIZE = 8;
        Screen screen = new Screen(SIZE);

        HuskyDog husky = new HuskyDog();
        husky.setPoint(3,6);             //设置的位置为第 3 行第 6 列
        screen.add(husky);
        screen.delay();
        husky.moveTo(4,6);               //移动到第 4 行第 6 列
    }
}
```

运行程序,"哈士奇"从屏幕中的第 3 行第 6 列移动到第 4 行第 6 列位置。

【例 5.3】 设计一个坐标 Point 类,包含两个成员变量 row、col,用于设置物体的位置信息。

Point 类有两个成员变量 row、col,接下来需要考虑该类需要哪些成员方法,首先需要设计方法设置和获得成员变量 row 和 col 的值,代码如下:

视频讲解

```java
public class Point{
    private int row;
    private int col;
    public void setPoint(int r, int c){
        row = r;
        col = c;
    }
    public int getRow(){
```

```
        return row;
    }
    public int getCol(){
        return col;
    }
}
```

有了 Point 类之后,可以将 HuskyDog 类的成员方法 moveTo()修改为:

```
public class HuskyDog{
    private int row;
    private int col;
    public void setPoint(int r, int c){
        row = r;
        col = c;
    }

    public void moveUp(){
        row -- ;
    }
    public void moveDown(){
        row ++ ;
    }
    public void moveLeft(){
        col -- ;
    }
    public void moveRight(){
        col ++ ;
    }

    public void moveTo(Point point){
        row = point.getRow();
        col = point.getCol();
    }
    public int getRow() {
        return row;
    }
    public int getCol() {
        return col;
    }
}
```

测试代码如下:

```
public class Main {
    public static void main(String[ ] args){
        final int SIZE = 8;
        Screen screen = new Screen(SIZE);

        HuskyDog husky = new HuskyDog();
        husky.setPoint(3,6);
        screen.add(husky);
```

```
        screen.delay();
        Point point = new Point();
        point.setPoint(5,6);
        
        husky.moveTo(point);
    }
}
```

运行程序,"哈士奇"从第 3 行第 6 列移动到了第 5 行第 6 列。

5.2.3 访问控制符

通过前面类的定义可知,无论是类本身,还是成员变量和成员方法都可以通过访问控制符控制访问权限。Java 语言提供了 4 种访问权限,控制级别从大到小依次为 public、protected、default 和 private,访问权限如表 5.1 所示。

表 5.1 访问权限

访 问 权 限	同一个类内部	同一个包内部	不同包中的子类	不同包的非子类
public	√	√	√	√
protected	√	√	√	×
default	√	√	×	×
private	√	×	×	×

1. 公共访问控制符 public

public 访问权限最宽松,如果一个类被 public 修饰,则该类为公共类,可被任何类访问。如果类的成员变量或成员方法被 public 修饰,则任何类都能够访问该成员变量或方法。

2. 保护访问控制符 protected

如果类的成员变量或成员方法被 protected 修饰,则同一个包里的类或者不同包中的子类可以访问该成员变量或方法。

3. 缺省访问控制符 default

如果一个类或者类的成员(包括成员变量和成员方法)没有使用任何访问修饰符修饰,说明它具有缺省访问控制特性。缺省访问控制权限规定该类只能被同一个包中的类访问,这种访问特性被称为包访问性。

4. 私有访问控制符 private

如果类的成员变量或成员方法被 private 修饰,则只有该类自身的成员可以访问该成员变量或方法。而其他任何类,包括该类的子类都不能访问。private 修饰非常重要,类的良好封装就是通过它来实现。

访问控制符能够有效实现访问权限的控制,如同发送微信朋友圈信息时设置访问权限一样,对于所有朋友都可以访问的信息,将其设置为公开;对于只能部分朋友访问的信息,则将其设置为部分可见。

通过设置访问权限,可以很好地对类进行封装,将对象的状态信息隐藏在对象内容中,实现信息隐藏。封装后不允许外部程序直接访问对象的内部信息,而是通过该类所提供的方法来实现对内部信息的操作访问。封装是面向对象的三大特征之一,是面向对象的核心思想。通过封装可以保护好内部信息,避免外界的干扰和保持类的独立性。

例如,HuskyDog 类的成员变量 row 和 col 的访问权限为私有访问权限,如果对象直接访问程序就会出错,代码如下:

```
HuskyDog husky = new HuskyDog();
husky.row = 25;
husky.col = 24;
```

运行程序,提示出相应的错误信息。

这样就避免直接访问数据导致出现各种风险情况,例如设计各种账户类时,如果能直接访问用户密码,后果不堪设想。在设计类的时候,通常情况下会将类的成员变量访问权限都设置为 private,这样可以很好地实现类的封装。

5.2.4 方法的重载

在设计类的成员方法时,会遇到这样一种情况:在同一个类中,需要设计一系列功能相似,但是应用场景有差别的方法。例如,设计一个计算类的求和方法,需要根据数字的个数和类型,设计多个求和方法适用于不同情况。如果每一个方法都用不同的方法名,使用时容易带来不便。Java 语言提供了方法重载机制,允许在一个类中定义多个同名的方法,这称之为方法重载。实现方法重载,要求同名的方法参数个数或者参数类型不同。例如,Screen 类的 add() 方法就使用了方法重载机制,参数既可以是单个的物体对象,也可以是数组。

视频讲解

【例 5.4】 编写程序,为 HuskyDog 类设计 moveTo() 方法,使得既可以通过坐标类 Point 类对象作为参数,也可以使用两个整型变量作为参数。

moveTo() 方法可以通过不同类型的参数实现相似的功能,这是典型的方法重载应用场景,代码如下:

```
public class HuskyDog{
    private int row;
    private int col;
    public void setPoint(int r, int c){
        row = r;
        col = c;
    }
    
    public void moveUp(){
        row -- ;
    }
    public void moveDown(){
        row ++ ;
    }
    public void moveLeft(){
        col -- ;
    }
    public void moveRight(){
        col ++ ;
    }
```

```java
    public void moveTo(int r,int c){
        row = r;
        col = c;
    }

    public void moveTo(Point point){
        row = point.getRow();
        col = point.getCol();
    }

    public int getRow() {
        return row;
    }
    public int getCol() {
        return col;
    }
}
```

测试代码如下:

```java
public class Main {
    public static void main(String[ ] args){
        final int SIZE = 8;
        Screen screen = new Screen(SIZE);

        HuskyDog husky = new HuskyDog();
        screen.add(husky);
        screen.delay();
        Point point = new Point();
        point.setPoint(5,6);
        husky.moveTo(point);

        screen.delay();
        husky.moveTo(7,6);
    }
}
```

运行程序,"哈士奇"在屏幕上不断地移动。

方法重载的优点是,功能类似的方法使用同一名字更容易记住,调用起来更简单。通过方法重载,可以实现编译时多态,编译器根据参数的不同调用相应的方法,具体调用哪个方法由编译器在编译阶段静态决定,所以也称之为静态多态。对于 C 语言较为熟悉的读者应该很快能够感受到方法重载的好处。需要注意的是,如果只有返回值不同,不能实现方法的重载。

5.2.5 构造方法

在上述设计的 HuskyDog 类中,每次实例化对象之后,都需要调用 setPoint()方法为位置属性赋值。如果想同 Screen 类一样,实例化对象的同时就为属性赋值,该如何实现?类的构造方法可以实现这种要求。构造方法是类的一种特殊的方法,作用是在创建对象时初始化对象状态。构造方法的定义格式如下:

```
[修饰符] 方法名([参数列表]){
    //方法体
}
```

构造方法的定义与普通方法定义格式相似,但区别是:
(1) 构造方法的名称必须与类名相同。
(2) 在方法名前没有返回值类型的声明。
熟悉了构造方法的定义格式后,通过一个案例了解构造方法的设计。

视频讲解

【例 5.5】 编写程序,为 HuskyDog 类设计构造方法。

构造方法与类同名,并且没有返回值类型说明,代码如下:

```java
public class HuskyDog{
    private int row;
    private int col;
    public HuskyDog(int r, int c){
        row = r;
        col = c;
    }

    public void moveUp(){
        row -- ;
    }
    public void moveDown(){
        row ++ ;
    }
    public void moveLeft(){
        col -- ;
    }
    public void moveRight(){
        col ++ ;
    }

    public int getRow() {
        return row;
    }
    public int getCol() {
        return col;
    }
}
```

测试代码如下:

```java
public class Main {
    public static void main(String[ ] args){
        final int SIZE = 8;
        Screen screen = new Screen(SIZE);

        HuskyDog husky = new HuskyDog(3,6);
        screen.add(husky);
        screen.delay();
        husky.moveUp();
    }
}
```

运行程序,"哈士奇"向上运动了一格。

每一个类至少有一个构造方法,在定义一个类的时候,若没有显式地定义构造方法,编译器将自动为类提供一个默认构造方法。

1. 默认构造方法

默认的构造方法没有参数,在其方法体中没有任何代码。如例 5.1 中 HuskyDog 类没有显式定义构造方法,编译器就提供了默认构造方法:

```
public HuskyDog(){
}
```

使用 new HuskyDog()创建 HuskyDog 类的对象时,系统自动调用了该类的默认构造方法。

2. 带参数的构造方法

由于系统自动提供的默认构造方法往往不能满足要求,因此可以在设计类的时候显式地定义构造函数。例如,HuskyDog 类中增加一个构造函数,通过构造函数初始化类的成员变量。需要注意的是,如果类定义了带参数的构造函数,编译器将不再提供默认的构造方法。例如,如果 HuskyDog 类只提供了带参数的构造函数,使用下列语句创建对象:

```
HuskyDog huskyDog = new HuskyDog ();
```

编译器会提示错误信息。

3. 构造方法的重载

如果想使用多种方式创建对象,则可以使用重载构造方法,通过这些重载的构造方法,在创建对象的时候可以灵活选择不同的构造方法,例如:

```
public class HuskyDog{
    private int row;
    private int col;
    public HuskyDog(){
    }

    public HuskyDog(int r, int c){
        row = r;
        col = c;
    }

    public HuskyDog(Point point){
        row = point.getRow();
        col = point.getCol();
    }

    public void moveUp(){
        row -- ;
    }
    public void moveDown(){
        row ++ ;
    }
    public void moveLeft(){
```

```
        col -- ;
    }
    public void moveRight(){
        col ++ ;
    }
    public int getRow() {
        return row;
    }
    public int getCol() {
        return col;
    }
}
```

测试代码如下：

```
public class Main {
    public static void main(String[ ] args){
        final int SIZE = 8;
        Screen screen = new Screen(SIZE);

        HuskyDog huskyA = new HuskyDog(3,6);
        screen.add(huskyA);

        Point point = new Point(5,5);
        HuskyDog huskyB = new HuskyDog(point);
        screen.add(huskyB);
    }
}
```

运行程序，屏幕上出现两只"哈士奇"。

4. this 关键字的使用

在 HuskyDog 类中使用成员变量 row 表示行信息，而在构造方法中使用参数 r 表示行信息，r 表示的信息不够明确，导致程序的可读性较差。如果将参数 r 也修改成 row，又会导致成员变量与局部变量同名冲突的问题，Java 语言中提供了 this 关键字可以很好地解决此问题。this 关键字表示对象本身，准确地说是对象的引用，使用方法如下：

this.属性名称

语句表示访问类中的成员变量，用来区分成员变量和方法参数重名问题。
例如：

```
public HuskyDog(int row, int col){
    this.row = row;
    this.col = col;
}
```

参数名与成员变量同名时，this.row 表示的是成员变量，而没带 this 的 row 表示的是方法参数。

this 关键字的另外一个用途是在一个构造方法中调用该类的另一个构造方法。例如，在

HuskyDog 类中定义无参构造方法调用了该类的另一个构造方法，代码如下：

```
public HuskyDog(){
    this(0,0);
}
```

使用 this 调用类的构造方法时，需要注意的是：该语句必须是第一条语句。
例如：

```
public HuskyDog(){
    int row = 0;
    int col = 0;
    this(row,col);
}
```

上述代码中，由于 this() 不是第一条语句，所以编译时会提示出现错误。
另外，只能在构造方法中使用 this 语句调用其他的构造方法，而不能在其他成员方法中使用。

5.2.6 static 关键字

前面完成的各种程序，如"贪吃蛇"的运动或者"俄罗斯方块"的运动中都忽视了越界的问题。解决越界问题的关键是需要知道屏幕大小。例如，HuskyDog 类的 MoveTo() 方法，若要保证不能运动到屏幕外，则需要根据屏幕的大小设置判断条件。在 HuskyDog 类中如何获得屏幕的大小？最直接的方法是新增一个参数，将 Screen 的对象引用传递进来，通过该对象获得屏幕的大小。这个方法虽然能解决问题，但是较为麻烦。Java 语言提供了 static 关键字，可以实现在没有创建任何对象的前提下，仅仅通过类本身来调用成员方法或者变量。

接下来通过一个例子来学习 static 关键字的作用。

【例 5.6】 设计一个 Config 配置类，保存游戏的基本信息。

游戏的配置信息包括游戏名称、版本信息、字体信息等基本信息，都可放置在配置类中，本例中只需要设置屏幕大小信息，代码如下：

视频讲解

```
public class Config {
    public static final int SCREENSIZE = 8;
}
```

测试代码如下：

```
public class Main {
    public static void main(String[ ] args){
        Screen screen = new Screen(Config.SCREENSIZE);
    }
}
```

运行程序，出现一个 8 行 8 列大小的屏幕。

static 关键字又称为静态修饰符，可以修饰类中的成员变量、方法以及修饰代码块优化程序性能。被 static 关键字修饰的变量或方法不需要依赖于对象来进行访问，可以通过类名直接去进行访问。

1. 静态变量

静态变量是被 static 关键字修饰的变量，这类变量属于类的变量，是一个公共的存储单元，被这个类所有的对象共享。任何一个对象操作它的时候，都是对同一个内存单元进行操作。例如，一个游戏的最高分，被所有玩家所共享，每一个玩家打破纪录，都是对同一数据进行操作。

视频讲解

【例 5.7】 设计一个 Player 玩家类，保存游戏玩家的 ID 信息。

每一款游戏都会有用户管理系统，管理游戏玩家的账号信息，如游戏 ID、用户名、账户密码等。本例设计的 Player 玩家信息类只需要设计游戏 ID，为了保证游戏 ID 唯一，ID 号为注册用户的总人数，代码如下：

```java
public class Player{
    public static int count = 0;
    private int id;
    public Player(){
        id = count;
        count++;
    }

    public int getID(){
        return id;
    }
}
```

测试代码如下：

```java
public class Main {
    public static void main(String[ ] args){
        final int SIZE = 8;
        Screen screen = new Screen(SIZE);

        Player player1 = new Player();
        Player player2 = new Player();

        System.out.println(Player.count);
    }
}
```

运行程序，在控制台上可以观察到结果为 2，意味着已经注册的用户数为 2。System 类是系统提供的系统类，该类定义了三个静态类变量，其中变量 out 是标准输出设备，通过 println() 方法输出信息。

2. 静态方法

静态方法是被 static 关键字修饰的方法，与静态变量一样，可以在不创建对象的情况下直接调用。例如，在 Config 类中增加静态方法获得屏幕的大小，代码如下：

```java
public class Config {
    public static final int SCREENSIZE = 8;
    public static int getScreenSize(){
        return SCREENSIZE;
    }
}
```

没有被 static 修饰的成员需要先创建对象,然后通过对象才能访问,而静态方法可以在不创建任何对象情况下调用,所以静态方法只能访问用 static 修饰的成员。

5.3 综合案例:重构"贪吃蛇"游戏

视频讲解

通过不断重构改善代码的设计,即使是很糟糕的代码,也可以改进成设计良好的代码。

Java 语言是面向对象的语言,接下来使用面向对象的思想重构"贪吃蛇"游戏。

编写程序,设计"贪吃蛇"类,实现按键"w""s""a""d"控制"贪吃蛇"上、下、左、右运行,如图 5.2 所示。

类的主要组成部分是成员变量和成员方法,设计类的时候需要先思考该类有哪些成员变量和方法。

1. 成员变量

"贪吃蛇"类需要的数据信息包括每个方块的位置、"贪吃蛇"的长度以及运动方向。所以"贪吃蛇"类的成员变量包括一个 Cell 类型的数组和两个 int 类型的变量,数组存储方块的位置信息,int 类型变量分别存储"贪吃蛇"的长度和运动方向。

2. 成员方法

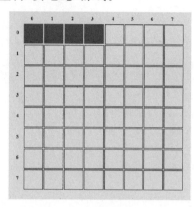

图 5.2 "贪吃蛇"游戏

成员方法包括控制"贪吃蛇"运动方向的方法、控制"贪吃蛇"运动的方法和获取方块的位置信息的方法。

根据分析,"贪吃蛇"类的代码如下:

```java
public class Snake {
    private Cell [ ] cells;
    private int len;               //存储蛇的长度
    private int dir;               //存储运动方向

    public Snake() {               //无参构造函数
        cells = new Cell[4];
        len = 4;
        dir = 2;                   //初始方向向下
        for(int i = 0; i < cells.length; i++ ) {
            cells[i] = new Cell(0,i);
        }
    }

    public Snake(int len, int dir) {   //有参构造函数
        cells = new Cell[len];
        this.len = len;
        this.dir = dir;
        for(int i = 0; i < cells.length; i++ ) {
            cells[i] = new Cell(0,i);
        }
    }
```

```java
    public Cell[ ] getCell(){
        return cells;
    }
    public void setDirection(int d) {            //设置运动方向
        dir = d;
    }

    public void move() {
        for(int i = len - 1; i > 0; i--) {       //蛇身运动
            cells[i].moveTo(cells[i-1]);
        }
        if(dir == 1) {                           //向上运动
            cells[0].moveUp();
        }
        if(dir == 2) {                           //向下运动
            cells[0].moveDown();
        }
        if(dir == 3) {                           //向左运动
            cells[0].moveLeft();
        }
        if(dir == 4) {                           //向右运动
            cells[0].moveRight();
        }
    }
}
```

测试代码如下:

```java
public class Main {
    public static void main(String[ ] args){
        final int SIZE = 8;
        Screen screen = new Screen(SIZE);
        Snake snake = new Snake();
        screen.add(snake.getCell());

        char key;
        while(true) {
            screen.delay();
            key = screen.getKey();
            if(key == 'w') {
                snake.setDirection(1);
            }
            if(key == 's') {
                snake.setDirection(2);
            }
            if(key == 'a') {
                snake.setDirection(3);
            }
            if(key == 'd') {
                snake.setDirection(4);
            }
```

```
            snake.move();
        }
    }
}
```

类的设计，往往不是一蹴而就，而是需要反复重构，才能使之结构清晰，具有灵活性以适应不断变化的需求。

例如，可以设计游戏 SnakeGame 类，成员变量包括"贪吃蛇"，成员方法包括初始化游戏、运行游戏，代码如下：

```
public class SnakeGame {
    private Snake snake;
    private Screen screen;
    public void init() {
        final int SIZE = 8;
        screen = new Screen(SIZE);
        snake = new Snake();
        screen.add(snake.getCell());
    }

    public void run() {
        char key = screen.getKey();

        while(true) {
            screen.delay();
            key = screen.getKey();
            if(key == 'w') {
                snake.setDirection(1);
            }
            if(key == 's') {
                snake.setDirection(2);
            }
            if(key == 'a') {
                snake.setDirection(3);
            }
            if(key == 'd') {
                snake.setDirection(4);
            }
            snake.move();
        }
    }
}
```

测试代码如下：

```
public class Main {
    public static void main(String[ ] args){
        SnakeGame game = new SnakeGame();
        game.init();
        game.run();
    }
}
```

设计一款新的游戏,比如"俄罗斯方块"游戏,会发现基本框架都与上面相似,先设计一个游戏类,然后设计游戏中的各种角色类。通过该案例可知,采用面向对象思维方式设计程序,更易扩展。

重构的每个步骤都很简单,比如修改变量的命名,或者删除多余的一句代码,修改一条语句。这些小改变看起来微不足道,但是聚沙成塔,累积起来就能形成质变,从根本上改善程序。

习题

5.1 以下关于构造方法描述正确的是_____。
   A. 构造方法的返回类型可以是 void 型
   B. 一个类只能拥有一个构造方法
   C. 构造方法是类的一种特殊方法,它的方法名必须与类名相同
   D. 构造方法的调用与其他方法相同

5.2 在类的定义中可以有两个同名的方法,这种面向对象程序特性称为_____。
   A. 封装　　　　B. 重载　　　　C. 继承　　　　D. 重写

5.3 构造方法的作用是_____。
   A. 访问类的成员变量　　　　B. 初始化成员变量
   C. 描述类的特征　　　　　　D. 保护成员变量

5.4 访问修饰符作用范围由大到小是_____。
   A. private-default-protected-public
   B. public-default-protected-private
   C. private-protected-default-public
   D. public-protected-default-private

5.5 以下对重载描述错误的是_____。
   A. 方法重载只能发生在一个类的内部　　B. 构造方法不能重载
   C. 重载要求方法名相同,参数列表不同　　D. 方法的重载与返回值类型无关

5.6 若一个无返回值无参数的方法 method() 是静态方法,则该方法正确的定义形式为_____。
   A. public void method(){}　　　　B. static void method(){}
   C. abstract void method(){}　　　D. void method(){}

5.7 Java 语言中,只限子类或者同一包中的类的方法能访问的访问权限是_____。
   A. public　　　　B. private
   C. protected　　 D. 无修饰

5.8 如果一个类的成员变量只能在所在类中使用,则该成员变量的修饰符必须是_____。
   A. public　　　　B. protected
   C. private　　　 D. static

5.9 设计"俄罗斯方块"某一种形状的类,如图 5.3 所示,实现按键控制其上、下、左、右运动,并编写测试程序。

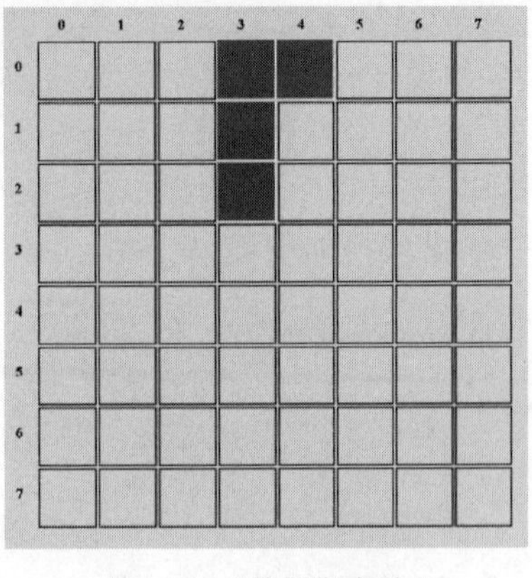

图 5.3 "俄罗斯方块"

第6章 面向对象特性

第5章介绍了类和对象的基本概念和使用方法,并且设计 Snake 和 SnakeGame 等类重构了"贪吃蛇"游戏,使游戏逻辑变得清晰。本章将继续利用面向对象的特性优化程序设计。

上一章介绍了面向对象的特征之一封装,面向对象还有两个重要的特性:继承和多态。继承是从已有的类中派生出新的类,在现有类的基础之上,增加新的方法或者重写已有的方法,产生新的类。继承可以使原有类得到重用,大大提高开发效率。多态是不同类的对象在调用同一个方法时所呈现出的多种不同行为。多态是在封装和继承的基础上体现出来,消除了类之间的耦合关系,使程序更加容易扩展。初学者在刚接触的时候会感到有些难度,本章会通过有趣的案例帮助读者掌握面向对象的特性。

6.1 类的继承

6.1.1 继承的概念

Java 语言中,类的继承是指从已有的类中派生新的类,构建出来的新类被称为子类,已经存在的类被称为父类、超类或基类。子类会自动拥有父类的属性和方法,同时也可以具有自己的特征。类的继承是创建新类的主要方法,这种方法可以使代码充分得到重用,有效地降低软件复杂性,提高开发效率,缩短开发周期。

例如,上一章的 HuskyDog 类和 TeddyDog 类有非常多相似的地方,可以设计一个 Dog 类,将两者的共性部分提出来,然后通过继承的方法实现代码的复用。实现类的继承关系使用 extends 关键字,格式如下:

```
[修饰符] class 子类名 extends 父类名{
    //类体定义
}
```

视频讲解

类的修饰符可选,用来指定类的访问权限。

【例6.1】 编写程序,实现Dog类作为父类,TeddyDog类和HuskyDog类作为子类继承Dog类。父类Dog类的代码如下:

```java
public class Dog {
    private int row;
    private int col;
    public Dog() {
    }

    public void setPoint(int r, int c){
        row = r;
        col = c;
    }

    public void moveUp(){
        row -- ;
    }
    public void moveDown(){
        row ++ ;
    }
    public void moveLeft(){
        col -- ;
    }
    public void moveRight(){
        col ++ ;
    }

    public int getRow() {
        return row;
    }
    public int getCol() {
        return col;
    }
}
```

"TeddyDog"和"HuskyDog"都继承于Dog类,代码如下:

```java
public class TeddyDog extends Dog{
}

public class HuskyDog extends Dog{
}
```

TeddyDog类和HuskyDog类继承于Dog类,就自动拥有了Dog类的成员属性和成员方法,测试代码如下:

```java
public class Main {
    public static void main(String[ ] args){
        final int SIZE = 8;
        Screen screen = new Screen(SIZE);
```

```
        HuskyDog husky = new HuskyDog();
        husky.setPoint(2, 2);
        screen.add(husky);

        TeddyDog teddy = new TeddyDog();
        teddy.setPoint(3, 6);
        screen.add(teddy);
    }
}
```

运行程序,屏幕上出现"哈士奇"和"泰迪"。

另外,还可以给新类增加新的特性。例如,给 HuskyDog 类新增向上跳(假设为向上运动两步)的方法,代码如下:

```
public class HuskyDog extends Dog{
    public void jump(){                    // 新增 jump 方法
        moveUp();
        moveUp();
    }
}
```

测试代码如下:

```
public class Main {
    public static void main(String[ ] args){
        final int SIZE = 8;
        Screen screen = new Screen(SIZE);

        HuskyDog husky = new HuskyDog();
        husky.setPoint(2, 2);
        screen.add(husky);

        screen.delay();
        husky.jump();
    }
}
```

通过上述例子可知,子类既可以继承父类的状态和行为,也可以具有自己的特征。Java 语言中的继承代表着真实世界中"is-a"关系(类的父子继承关系)。例如,哈士奇是狗的一种。

在 Java 语言中使用继承的时候,需要注意:

(1) 类仅支持单继承,不支持多重继承,即一个类只能有一个直接父类。

(2) 类支持多层继承,即一个类的父类也可以继承于其他类。

6.1.2　方法重写

在继承关系中,子类会自动继承父类中的方法,但是有时候,需要对继承的方法进行修改,这就是方法的重写或者称为方法的覆盖。

例如,游戏道具"哈士奇"HuskyDogToy 类也继承于 Dog 类,但是道具类"哈士奇"在屏幕中不会上、下、左、右移动,所以 HuskyDogToy 类需要重写上、下、左、右运动的方法,代码如下:

```java
public class HuskyDogToy extends Dog{
    public void moveUp() {
    }
    public void moveDown() {
    }
    public void moveLeft() {
    }
    public void moveRight() {
    }
}
```

测试代码如下:

```java
public class Main {
    public static void main(String[ ] args){
        final int SIZE = 8;
        Screen screen = new Screen(SIZE);

        HuskyDogToy husky = new HuskyDogToy(2,2);
        screen.add(husky);

        screen.delay();
        husky.moveUp();
    }
}
```

运行程序,进行向上运动的操作后,游戏道具"哈士奇"仍然保持原地不动。从运行结果来看,调用 HuskyDogToy 类的 moveUp()方法时,只调用了子类重写的方法,并不会调用父类的 moveUp()方法。

子类重写父类方法时需要注意的事项如下:

(1) 父类访问权限为 private 的方法不能被子类重写。

(2) 子类重写方法时不能降低方法的访问权限。例如,父类中的方法访问权限是 public,子类重写该方法的访问权限就不能是 private。

6.1.3　super 关键字的使用

当子类重写父类的方法后,子类对象将无法直接访问父类被重写的方法。程序设计时,有时候需要在子类中调用父类被重写的方法,为了应对此类场景,Java 语言提供了 super 关键字用来访问父类的成员。super 关键字可用于如下三种情况:

(1) 在子类中调用父类被重写的方法。

(2) 在子类中访问父类被隐藏的成员变量。

(3) 在子类中调用父类的构造方法。

视频讲解

【例 6.2】　设计 HuskyDog 类,它继承于 Dog 类,但是 HuskyDog 类的上、下、左、右移动速度是普通 Dog 类的两倍。

本例中,HuskyDog 类继承于 Dog 类,但是它的移动速度是普通 Dog 类的两倍,所以需要重写上、下、左、右移动方法,并且还要调用父类被重写的方法,代码如下:

```java
public class HuskyDog extends Dog{
    public void moveUp() {
        super.moveUp();
        super.moveUp();
    }

    public void moveDown() {
        super.moveDown();
        super.moveDown();
    }

    public void moveLeft() {
        super.moveLeft();
        super.moveLeft();
    }

    public void moveRight() {
        super.moveRight();
        super.moveRight();
    }
}
```

测试代码如下:

```java
public class Main {
    public static void main(String[ ] args){
        final int SIZE = 8;
        Screen screen = new Screen(SIZE);

        HuskyDog husky = new HuskyDog();
        husky.setPoint(2, 2);
        screen.add(husky);

        screen.delay();
        husky.moveUp();
    }
}
```

运行程序,"哈士奇"每次运动速度都是普通类型狗的 2 倍。

6.1.4 子类的构造方法及调用过程

Java 语言规定,在创建子类对象时,会先创建该类的所有父类对象。编写子类的构造方法时,都会先调用父类的构造方法。

子类构造方法中调用父类的构造方法有下列两种。

(1) 调用父类的默认构造方法。

在子类构造方法中若没有显式使用 super()方法调用父类的构造方法,则编译器将在子类构造方法的第一句自动加上 super()方法,即调用父类无参数的构造方法。

例如:

```
public class TeddyDog extends Dog{
    public TeddDog(){
    }
}
```

等同于:

```
public class TeddyDog extends Dog{
    public TeddDog(){
        super();
    }
}
```

TeddyDog 类的构造方法被调用时,执行了 super()方法,从而调用了父类的默认构造方法。需要注意的是,通过 super()方法调用父类构造方法的代码必须位于子类构造方法的第一行,并且只能出现一次。

(2) 调用父类的有参构造方法,具体方法如下:

```
super(参数列表);
```

例如,Dog 类提供了有参构造函数,代码如下:

```
public class Dog {
    private int row;
    private int col;
    public Dog(int row, int col) {
        this.row = r;
        this.col = c;
    }

    public void setLocation(int row, int col) {
        this.row = row;
        this.col = col;
    }

    public int getRow() {
        return row;
    }
    public int getCol() {
        return col;
    }

    public void moveUp() {
        row -- ;
    }
    public void moveDown() {
        row ++ ;
    }
    public void moveLeft() {
        col -- ;
    }
    public void moveRight() {
```

```
            col ++ ;
        }
    }
```

因为 Dog 类显式提供了有参的构造方法,而没有提供无参的构造方法,则 TeddyDog 类继承 Dog 类时也需要显式提供有参构造方法,并且通过 super 语句调用父类的有参构造方法,代码如下:

```
public class TeddyDog extends Dog{
    public TeddDog(int row, int col){
        super(row,col);
    }
}
```

测试代码如下:

```
public class Main {
    public static void main(String[ ] args){
        final int SIZE = 8;
        Screen screen = new Screen(SIZE);

        TeddyDog teddy = new TeddyDog (2,2);
        screen.add(teddy);

        screen.delay();
        teddy.moveUp();
    }
}
```

运行程序,"泰迪"向上运动一格。

6.1.5 final 修饰符

继承是面向对象的三大特性之一,是创造新类的主要方法,但是有时也需要对继承进行限制。例如,设计密码管理类或者数据库信息管理类等,为了安全考虑,不允许被其他类继承。Java 语言提供了 final 关键字修饰类,它有"不可改变"或者"终态"含义。被 final 关键字修饰的类不能被继承,没有子类。final 修饰符除了修饰类外,还可以修饰方法和变量。

1. final 修饰变量

用 final 修饰的变量包括方法的参数、局部变量和类的成员变量。被 final 修饰的变量,一旦被赋值,其值不能改变,故称其为常值变量。如果再次对该变量进行赋值,会产生编译错误。例如,定义屏幕大小:

```
final int SIZE = 8;
```

重新对变量 SIZE 赋值,如:

```
SIZE = 16;
```

程序在编译时会提示相应的错误信息。

2. final 修饰方法

如果一个方法被 final 修饰,则该方法可以被子类继承,但是不能被子类重写。例如,将 Dog 类中 moveUp()方法设置为 final 方法,代码如下:

```
public final void moveUp() {
    row--;
}
```

如果 TeddyDog 类继承 Dog 类,则不能重写 moveUp()方法,代码如下:

```
public class TeddyDog extends Dog{
    public void moveUp() {
        super.moveUp();
        super.moveUp();
    }
}
```

程序在编译时会提示相应的错误信息。

当父类中定义的某个方法不希望被子类重写时,就可以用 final 修饰该方法,使之成为最终方法。

3. final 修饰类

类被 final 修饰后,则该类为最终类,不会被继承。例如:

```
final class Dog{
    //方法体
}
```

如果 TeddyDog 类继承于 Dog 类,如:

```
public class TeddyDog extends Dog{
}
```

程序在编译时会提示相应的错误信息。

有时候为了安全考虑,防止类被继承,可以使用 final 修饰符。在 Java 系统提供的库中有一些类为 final 类,如 Math 类和 String 类,都是最终类,不能被继承。

6.1.6 Object 类

Object 类是所有类的父类,因此 Object 类也被称为根类。定义类时如果没有使用 extends 关键字显式地指定父类,编译器会自动加上"extends Object"。Object 类的常用方法如表 6.1 所示,这些方法都会被子类继承。

表 6.1 Object 类的常用方法

方法	功能
String toString()	返回对象的字符串表示形式
boolean equals(Object obj)	比较两个对象是否相等
int hashCode()	返回对象的散列码值

1. toString()方法

toString()方法是 Object 类中的常用方法,该方法返回一个字符串,字符串由类名加上一个@符号和十六进制数表示的散列码值组成。在实际开发中,通常希望 toString()方法包含更多有用信息,因此需要重写 toString()方法。

2. equals()方法

equals()方法用来比较两个对象是否相等,Object 类中的 equals()方法用来判断调用 equals()方法的对象和形参 obj 所引用的对象是否为同一对象。所谓同一对象就是指内存中同一块存储单元,即便是内容完全相等的对象,但是存储在不同的内存单元,返回结果也是 false。在实际开发中,经常需要比较两个对象的内容是否相等,因此也需要重写 equals()方法。

3. hashCode()方法

hashCode()方法返回对象的散列码值,该散列码值就是对象在计算机内部存储单元地址的十进制表示形式。

6.2 抽象类和接口

6.2.1 抽象类

前面设计的 HuskyDog 类和 TeddyDog 类,它们都继承于 Dog 类。根据 HuskyDog 类和 TeddyDog 类可以分别生成具体的对象,但是使用 Dog 类生成具体对象的时候,会产生疑问:究竟生成什么狗的对象?此时,Dog 类是一个抽象概念。如果一个类中没有足够的信息来描述一个具体的对象,它就是抽象类。

定义抽象类的基本格式如下:

```
[修饰符] abstract class 类名{
}
```

抽象类必须使用 abstract 关键字来修饰,不能进行实例化,也就是不能用来创建对象,但是可以定义抽象方法。设计类的时候,把共性的行为提取到父类中,又无法准确对其进行描述,并且这个行为还是子类必须要实现的行为,就可以将其定义为抽象方法。

例如,在上一章设计 SnakeGame 类的时候,可以发现:不同游戏的基本框架是一样的,都包含初始化游戏和运行游戏两部分。因此,可以设计一个游戏框架类,代码如下:

```
public Class GameCore{
    public void init(){
        //初始化游戏精灵
    }
    public void run(){
        while(true){
            //更新游戏精灵
        }
    }
}
```

有了这个框架类之后，设计新的游戏时只需要按照既定的框架完成两部分内容，即初始化游戏精灵和更新游戏精灵，这使得设计新的游戏变得简单。新的游戏框架类只需继承 GameCore 类，然后重写初始化游戏精灵方法和更新游戏精灵方法。这两个方法在子类必须实现，并且不同的子类实现不同的方法，所以可以将其设计成抽象方法。

抽象方法的定义必须用 abstract 关键字来修饰，并且在定义方法时不需要实现方法体，具体格式如下：

```
[修饰符] abstract 返回值类型 方法名([参数列表]);
```

将 GameCore 类设计成抽象类，并为其设计两个抽象方法，代码如下：

```java
public abstract Class GameCore{
    public void init(){
        initSprite();
    }

    public void run(){
        while(true){
            update();
        }
    }

    public abstract void initSprite();
    public abstract void update();
}
```

抽象类不能实例化，如果想利用抽象类创建对象，必须创建子类重写抽象父类中所有抽象方法。如果子类没有重写父类的全部抽象方法，那么子类也是抽象类，也不能实例化。

例如，利用 GameCore 抽象类重构 SnakeGame 类，代码如下：

```java
public class SnakeGame extends GameCore{
    private Snake snake;
    private Screen screen;

    @Override
    public void initSprite() {
        screen = new Screen(Config.SCREENSIZE);
        snake = new Snake();
        Cell [] snakeCell = snake.getCell();
        for(int i = 0; i < snake.getLen();i++) {
            screen.add(snakeCell[i]);
        }
    }

    @Override
    public void update() {
        char key = screen.getKey();
        screen.delay();
        key = screen.getKey();
        if(key == 'w') {
```

```
            snake.setDirection(1);
        }
        if(key == 's') {
            snake.setDirection(2);
        }
        if(key == 'a') {
            snake.setDirection(3);
        }
        if(key == 'd') {
            snake.setDirection(4);
        }
        snake.move();
    }
}
```

"@Override"用于表示被标注的方法是一个重写方法,去掉也不影响程序的执行。使用@Override注解的作用是提醒这是一个重写方法,提高代码的可读性,帮助检查是否正确地重写父类的方法。

测试代码如下:

```
public class Main {
    public static void main(String[ ] args){
        SnakeGame game = new SnakeGame();
        game.init();
        game.run();
    }
}
```

运行程序,"贪吃蛇"游戏能够正常运行。通过案例可知,子类全部实现了父类的抽象方法后,可以正常实例化创建对象。

抽象类能让共性的部分实现代码复用,避免重复设计,又充分允许各自实现个性部分。比如设计婚礼流程类时,将其设计成抽象类,既能保证婚礼按照固定流程进行,又允许每个人的婚礼充满独特元素。抽象类的用途非常广泛,能使程序变得灵活,在本书网络版游戏中网络通信部分也采用了抽象类设计思想。

使用抽象类和抽象方法时需要注意的是,包含抽象方法的类必须为抽象类,但是抽象类可以不包含任何抽象方法。例如,将上述Dog类设计成抽象类,却不用包含任何抽象方法。

6.2.2 接口

接口是一种特殊的抽象类,设计初衷是对行为进行抽象,用于定义一套标准、约束与规范等。在Java 8以前,接口是常量和方法的集合,方法只能是抽象方法,没有具体的实现。在Java 8中,将接口做了补充,支持default定义的普通方法与静态方法,并且支持Lambda表达式,用清晰简洁的表达式表达一个接口。

抽象类与接口的区别在于抽象类是对类的抽象,而接口是对行为的抽象,接口的重要作用是实现多重继承。Java语言中仅支持单重继承,不支持多重继承,即一个子类只能有一个直接父类。但是实际操作中,类之间的继承关系往往是多重继承的关系。为了实现多重继承,Java提供了接口机制,使不相关的类也可以具有相同的行为。

例如，在经典的先秦古籍《山海经》里面，记载着许多极具神秘感的奇异怪兽，比如凤凰、朱雀、玄武和穷奇等，其中穷奇是一种会飞的老虎。该如何设计穷奇类？首先设计一个老虎类，因为《山海经》中还有其他5种老虎，这样使得代码可以得到复用。然后穷奇类继承于老虎类，新增一个飞翔的方法。但是，凤凰、朱雀也能飞翔，它们又不属于老虎类，于是将飞翔这一行为设计成接口，具备飞翔这一行为的类都可以实现该接口。

接口的定义与类的定义类似，包括接口声明和接口体两部分，定义接口是 interface 关键字，一般格式如下：

```
[修饰符] interface 接口名[extends 父接口1,父接口2…]{
    //接口体
}
```

extends 表示该接口继承了哪些接口，一个接口可以继承多个接口，这解决了类的单继承限制。

接口体包含常量定义和方法定义两部分，常量的定义可以省略修饰符，此时系统会默认加上"pubic static final"属性。接口在定义抽象方法时，也可以省略修饰符，系统默认加上"public abstract"修饰符。

接口的抽象方法只能通过类实现，因此需要定义一个接口的实现类，该类通过 implements 关键字实现接口，并实现接口中所有抽象方法。定义接口的实现类语法格式如下：

```
[修饰符] class 类名[extends 父类名] [implements 接口1,接口2…]{
    // 类体定义
}
```

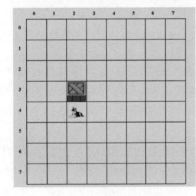

图6.1 "推箱子"游戏

视频讲解

一个类可以实现多个接口，需要在 implements 子句中对实现的接口用逗号隔离。如果把接口理解成特殊的类，那么类利用接口就实现了多重继承。

【例6.3】 编写程序，实现"哈士奇"推"箱子"的小游戏功能，如图6.1所示。

要实现"哈士奇"推"箱子"的小游戏，首先需要设计两个类："哈士奇"类和"箱子"类，"箱子"在"哈士奇"的作用下，也会上、下、左、右运动，所以两个类都有上、下、左、右运动的方法。设计的时候，既可以将上、下、左、右运动设置为一个类，然后让"哈士奇"类和"箱子"类去继承，也可以将上、下、左、右运动设置为接口，让两个类分别去实现接口。使用接口的方式更加灵活，本例将上、下、左、右运动的方法设置为接口，代码如下：

```java
public interface Move {
    void moveUp();
    void moveDown();
    void moveLeft();
    void moveRight();
}
```

为了使设计简单，项目提供了 GameObject 类，可以非常方便地在屏幕上添加各种游戏精灵。

GameObject 类的代码如下：

```java
public class GameObject {
    private int row;
    private int col;
    private int size;                    //游戏精灵的大小
    private boolean life;                //游戏精灵的生命值

    public GameObject() {
        life = true;
        size = 1;
    }

    public GameObject(int r, int c) {
        row = r;
        col = c;
        life = true;
        size = 1;
    }

    public int getRow() {
        return row;
    }
    public void setRow(int row) {
        this.row = row;
    }

    public int getCol() {
        return col;
    }
    public void setCol(int col) {
        this.col = col;
    }

    public int getSize() {
        return size;
    }
    public void setSize(int size) {
        this.size = size;
    }

    public boolean getLife() {
        return life;
    }
    public void setLife(boolean life) {
        this.life = life;
    }
}
```

通过继承 GameObject 类，很容易创建各种游戏精灵。需要注意的是，为了能在屏幕上正常显示游戏精灵，读者自己设计游戏精灵类的时候，除了继承 GameObject 类外，还需要在项目的 Images 文件夹下建立与之对应的图像文件，图像文件名要与类名相同，比如 HuskyDog 类对应的图像文件为"HuskyDog.png"。

"哈士奇"类的代码如下：

```java
public class HuskyDog extends GameObject implements Move{
    public HuskyDog(int row, int col) {
        super(row,col);
    }

    public void moveUp() {
        setRow(getRow() - 1);
    }
    public void moveDown() {
        setRow(getRow() + 1);
    }
    public void moveLeft() {
        setCol(getCol() - 1);
    }
    public void moveRight() {
        setCol(getCol() + 1);
    }


    public boolean isHit(Box box) {                    // 判断是否撞上了箱子
        if(box.getRow() == getRow() && box.getCol() == getCol()) {
            return true;
        }
        else {
            return false;
        }
    }
}
```

"箱子"类的代码如下：

```java
public class Box extends GameObject implements Move{
    public Box(int row, int col) {
        super(row,col);
    }

    public void moveUp() {
        setRow(getRow() - 1);
    }
    public void moveDown() {
        setRow(getRow() + 1);
    }
    public void moveLeft() {
        setCol(getCol() - 1);
    }
    public void moveRight() {
        setCol(getCol() + 1);
    }
}
```

设计好了游戏需要的精灵类之后，接下来设计游戏核心类 SokobanGame，继承于抽象类 GameCore，需要实现两个抽象方法 initSprite() 和 update()，代码如下：

```java
public class SokobanGame extends GameCore{
    private HuskyDog husky;
    private Box box;
    private Screen screen;

    public void initSprite() {
        final int SIZE = 8;
        screen = new Screen(SIZE);

        husky = new HuskyDog(4,2);
        box = new Box(3,2);
        screen.add(husky);
        screen.add(box);
    }

    public void update() {
        char key = screen.getKey();
        screen.delay();
        key = screen.getKey();
        if(key == 'w') {
            husky.moveUp();
            if(husky.isHit(box)) {
                box.moveUp();
            }
        }

        if(key == 's') {
            husky.moveDown();
            if(husky.isHit(box)) {
                box.moveDown();
            }
        }

        if(key == 'a') {
            husky.moveLeft();
            if(husky.isHit(box)) {
                box.moveLeft();
            }
        }

        if(key == 'd') {
            husky.moveRight();
            if(husky.isHit(box)) {
                box.moveRight();
            }
        }
    }
}
```

测试代码如下:

```java
public class Main {
    public static void main(String[ ] args){
        SokobanGame game = new SokobanGame();
        game.init();
        game.run();
    }
}
```

运行程序,通过按键"w""s""a""d"可以实现"哈士奇"推着"箱子"上、下、左、右运动。通过这个例子可以感受到抽象类和接口的优势。

在 Java 8 之前的版本,接口要求只能有抽象方法。从 Java 8 之后的版本开始,接口除了抽象方法外,还可以有默认方法和静态方法。默认方法要在返回值前加上修饰符 default,可以通过接口实现类的对象的调用,静态方法可以直接用接口名进行调用。Java 9 版本中接口不仅支持默认方法,还支持私有方法和私有静态方法。例如,许多动物都有叫喊行为,对于这种在多个类中都包含的行为,可以将其设置为接口,代码如下:

```java
public interface Sound {
    public default void play(String filename){
        try{
            Clip clip = AudioSystem.getClip();
            clip.open(AudioSystem.getAudioInputStream(new File(filename)));
            clip.start();
        }
        catch (Exception e){
            e.printStackTrace(System.out);
        }
    }
}
```

上述代码中 play()方法作用是播放音频文件,输入的参数是音频文件的文件名,通过设置参数可以播放不同的音频文件。对于 play()方法中的具体代码可以无须了解,只需要了解其作用即可。

通过 Sound 接口,可以为"哈士奇"类新增叫喊的功能,代码如下:

```java
public class HuskyDog extends GameObject implements Move,Sound{
    public HuskyDog(int row, int col) {
        super(row, col);
    }

    @Override
    public void moveUp() {
        setRow(getRow() - 1);
    }

    @Override
    public void moveDown() {
        setRow(getRow() + 1);
    }

    @Override
```

```java
    public void moveLeft() {
        setCol(getCol() - 1);
    }

    @Override
    public void moveRight() {
        setCol(getCol() + 1);
    }

    public void play() {
        String s = "sound/HuskyDog.wav";
        play(s);
    }
}
```

测试代码如下:

```java
public class Main {
    public static void main(String[] args){
        final int SIZE = 8;
        Screen screen = new Screen(SIZE);

        HuskyDog husky = new HuskyDog(6,2);
        screen.add(husky);

        String s = "sound/HuskyDog.wav";
        husky.play(s);
    }
}
```

运行程序,屏幕上出现了小狗,并且可以听到小狗的叫声。

也可以将 Sound 接口中的 play() 方法改为静态方法,代码如下:

```java
public interface Sound {
    public static void play(String filename){
        try{
            Clip clip = AudioSystem.getClip();
            clip.open(AudioSystem.getAudioInputStream(new File(filename)));
            clip.start();
        }
        catch (Exception e){
            e.printStackTrace(System.out);
        }
    }
}
```

在"哈士奇"类中新增一个 play() 方法,直接调用接口的静态方法,代码如下:

```java
public class HuskyDog extends GameObject implements Move{
    public HuskyDog(int row, int col) {
        super(row, col);
    }
```

```java
    @Override
    public void moveUp() {
        setRow(getRow() - 1);
    }

    @Override
    public void moveDown() {
        setRow(getRow() + 1);
    }

    @Override
    public void moveLeft() {
        setCol(getCol() - 1);
    }

    @Override
    public void moveRight() {
        setCol(getCol() + 1);
    }

    public void play() {
        String s = "sound/HuskyDog.wav";
        Sound.play(s);
    }
}
```

测试代码如下：

```java
public class Main {
    public static void main(String[ ] args){
        final int SIZE = 8;
        Screen screen = new Screen(SIZE);

        HuskyDog husky = new HuskyDog(6,2);
        screen.add(husky);
        husky.play();
    }
}
```

运行程序，屏幕上出现了小狗，并且可以听到小狗的叫声。

无论是程序设计，还是现实生活中，接口都是一个非常有用的思想。接口定义了一种标准规范，有了标准规范之后，大家都知道需要做哪些事，但是每件事的具体实现可以采用不同方式。接口使得设计与实现分离。例如，大部分人在结婚的流程上需要完成的事情都一样，但是每一件事完成的方式可以不同。再如 USB 接口，不同厂家的 USB 接口实现的方法各不相同，但是它们都遵循统一的标准，可以轻易实现互联互通。这就是标准的意义，节省了学习、沟通和交流的成本。

6.3 多态

6.3.1 多态概述

多态是指同一个行为具有多个不同表现形式或形态。比如，在经典的多人在线竞技游戏中，不同的"英雄"有着不同的技能,执行释放技能这个行为的时候,会根据具体英雄产生不同效果。Java语言中的多态分为两种:一种是编译期的多态,另外一种是运行期的多态。编译期多态通过方法重载实现,运行期多态是指不同类的对象在调用同一个方法时呈现的多种不同行为,简而言之就是同一行为发生在不同对象中产生不同结果。

多态的好处就是减少了类之间的耦合关系,增强了可替换性,提高了程序的可维护性和可扩展性。例如,Screen 类的 add()方法里就使用了多态,在屏幕里新增一种物体非常容易,不需要修改 Screen 类里的任何代码。

多态使用的方式就是通过父类的引用指向子类对象,程序运行时通过动态绑定来实现对子类方法的调用,这就是多态性。

【例 6.4】 编写程序,实现"哈士奇""泰迪"的攻击技能,"哈士奇"攻击技能是叫喊,"泰迪"攻击技能是变大。

父类 Dog 类的代码如下:

```java
public abstract class Dog extends GameObject implements Move{
    public Dog(int r, int c) {
        super(r,c);
    }

    @Override
    public void moveUp() {
        setRow(getRow() - 1);
    }

    @Override
    public void moveDown() {
        setRow(getRow() + 1);
    }

    @Override
    public void moveLeft() {
        setCol(getCol() - 1);
    }

    @Override
    public void moveRight() {
        setCol(getCol() + 1);
    }

    public abstract void attack();
}
```

HuskyDog 继承了 Dog 类,并重写了 attack 方法,代码如下:

```java
public class HuskyDog extends Dog {
    public HuskyDog(int r, int c) {
        super(r, c);
    }

    public void play() {
        String s = "sound/HuskyDog.wav";
```

```
        Sound.play(s);
    }

    @Override
    public void attack() {
        play();
    }
}
```

TeddyDog 类也继承了 Dog 类,并重写了 attack 方法,代码如下:

```
public class TeddyDog extends Dog{
    public TeddyDog(int r, int c) {
        super(r, c);
    }

    @Override
    public void attack() {
        setSize(2);
    }
}
```

测试代码如下:

```
public class Main {
    public static void main(String[ ] args){
        final int SIZE = 8;
        Screen screen = new Screen(SIZE);

        Dog husky = new HuskyDog(6,2);
        screen.add(husky);

        Dog teddy = new TeddyDog(2,5);
        screen.add(teddy);

        screen.delay();
        husky.attack();
        teddy.attack();
    }
}
```

运行程序,"哈士奇"叫喊,"泰迪"变大 4 倍。通过上述例子可以了解多态的使用方式,但是没有明显看到多态的优势。接下来将例题稍作修改就能发现多态的优势。设计一个方法完成两只狗相互攻击的功能。两只狗相互打架的可能性有许多种:两只哈士奇相互之间打架、两只泰迪之间打架、一只泰迪和一只哈士奇打架等。面对这些情况,如果每一种情况设计一个方法的话,就会变得非常复杂,而使用多态就能够很容易解决该问题。为 Dog 类新增一个 fight()方法,代码如下:

```
public abstract class Dog extends GameObject implements Move{
    public Dog(int r, int c) {
        super(r,c);
    }
```

```java
    @Override
    public void moveUp() {
        setRow(getRow() - 1);
    }

    @Override
    public void moveDown() {
        setRow(getRow() + 1);
    }

    @Override
    public void moveLeft() {
        setCol(getCol() - 1);
    }

    @Override
    public void moveRight() {
        setCol(getCol() + 1);
    }

    public void fight(Dog dog){
        attack();
        dog.attck();
    }
    public abstract void attack();
}
```

而 HuskyDog 类和 TeddyDog 类不需要改变，测试代码如下：

```java
public class Main {
    public static void main(String[ ] args){
        final int SIZE = 8;
        Screen screen = new Screen(SIZE);

        Dog husky = new HuskyDog(6,2);
        screen.add(husky);

        Dog teddy = new TeddyDog(6,4);
        screen.add(teddy);

        screen.delay();
        husky.fight(teddy);
    }
}
```

运行程序，"哈士奇"与"泰迪"打架，"哈士奇"会叫喊，而"泰迪"则会变大。读者可以修改测试代码，测试"泰迪"与"泰迪"或者"哈士奇"与"哈士奇"相互打架的情况，会发现上述代码能够很好地适用于这些情况。这就是多态的优势，如果不使用多态的话，对于不同种类的狗之间相互打架，就需要根据情况设计多个方法，而不是像上述代码一样，只需要一个 fight() 方法就能解决各种情况。

6.3.2 对象的类型转换

多态使用中,由于子类继承了父类的属性和行为,因此子类对象可以作为父类对象使用,即子类对象可以自动转换为父类对象,这种转换被称为"向上转型"。例如:

```
Dog husky = new HuskyDog();    //将 HuskyDog 类对象当作 Dog 类型来使用
```

将子类对象当作父类对象使用的时候,不需要进行任何显式声明,子类对象可以直接调用父类的方法。但是需要注意的是,此时不能通过父类对象去调用子类特有的方法。例如,play()方法是 HuskyDog 类特有的方法,而父类 Dog 类并没有该方法,如果使用如下代码则会出现编译错误:

```
Dog husky = new HuskyDog();
husky.play();
```

程序编译的时候会将 husky 对象当作 Dog 类对象,而在运行期才会根据对象具体类型调用相应的方法,这就是运行期绑定。由于 Dog 类中没有 play()方法,所以编译时会提示错误信息。

如果父类对象想使用子类对象特有的方法,则需要将其转换成子类对象,这种转换被称为"向下转型",这时需要使用强制类型转换,语法格式如下:

```
子类类型 引用名 = (子类类型)父类引用;
```

例如:

```
Dog dog = new HuskyDog();
HuskyDog husky = (HuskyDog) dog;
husky.play();
```

向上转换可以将任何对象转化为继承链中任何一个父类对象。而向下转换,则要求父类对象是用子类构造方法生成的,这样的转换才正确。例如:

```
Dog dog = new TeddyDog();
HuskyDog husky = (HuskyDog) dog;
```

这种转换就不正确,转换只发生在有继承关系的类和接口之间。向下转换有风险,所以在转换的时候,需要使用 instanceof 运算符来判断一个对象是否为某个类(或接口)的实例或者是其子类实例,语法格式如下:

```
对象 instanceof 类(或接口)
```

返回数据类型为 boolean 型,如果对象是该类的对象,返回"true",否则返回"false"。例如:

```
Dog dog = new TeddyDog();
if(dog instanceof HuskyDog){
    HuskyDog husky = (HuskyDog) dog;
}
```

instanceof 运算符判断对象 dog 是否为 HuskyDog 类型,如果是 HuskyDog 类型,则强制转换为 HuskyDog 类型。

Java 17 中新增的特性,可以将上述代码简写,将类型转换和变量声明都在 if 中处理并且直接使用这个变量,代码如下:

```
if(dog instanceof HuskyDog husky){
}
```

6.3.3 接口实现多态

通过接口也可以实现多态,接口是一种引用类型,任何实现该接口的实例都可以存储在该接口的变量中。当通过接口对象调用某个方法时,运行时系统确定该调用哪个类的方法,利用这一特性,接口也能实现多态。

例如,HuskyDog 类和 TeddyDog 类都实现了 Move 接口,而接口实例可以指向不同的实现对象,测试代码如下:

```
public class Main {
    public static void main(String[ ] args){
        final int SIZE = 8;
        Screen screen = new Screen(SIZE);

        HuskyDog husky = new HuskyDog(6,2);
        screen.add(husky);

        TeddyDog teddy = new TeddyDog(6,5);
        screen.add(teddy);

        Move huskyMove = husky;
        Move teddyMove = teddy;

        screen.delay();
        huskyMove.moveUp();
        teddyMove.moveDown();
    }
}
```

运行程序,"哈士奇"向上运动,"泰迪"向下运动。观察代码可以发现,任何实现该接口的实例都可以存储在该接口的变量中,这使得利用接口也可以很容易实现多态。

【例 6.5】 编写程序,实现鼠标控制"哈士奇"释放技能。
Java 图形库提供了鼠标监听 MouseListener 接口,该接口包含 5 个方法。

(1) void mouseClicked(MouseEvent e):在组件上单击(按下并释放)鼠标按钮时调用。
(2) void mouseEntered(MouseEvent e):鼠标进入组件时调用。
(3) void mouseExited(MouseEvent e):鼠标退出组件时调用。
(4) void mousePressed(MouseEvent e):在组件上按下鼠标按钮时调用。
(5) void mouseReleased(MouseEvent e):在组件上释放鼠标按钮时调用。

当按下、释放或单击(按下并释放)鼠标时会生成鼠标事件,鼠标光标进入或离开组件时也会生成鼠标事件。发生鼠标事件时,将调用监听器对象中的相应方法,并将 MouseEvent 事件传递给该方法。

项目中 Screen 类提供了 addMouseListener(MouseListener l)方法将创建的鼠标监听对象向

视频讲解

该屏幕组件注册。addMouseListener()方法的参数是 MouseListener 接口类型的变量,则任何实现该接口的类产生的实例都可以作为参数。例如,设计 MouseAttack 类实现了鼠标监听 MouseListener 接口,该类的实例对象可以作为 addMouseListener()方法的参数,代码如下:

```java
public class MouseAttack implements MouseListener{
    Dog dog;
    public MouseAttack(Dog dog) {
        this.dog = dog;
    }

    @Override
    public void mouseClicked(MouseEvent e) {
        dog.attack();
    }

    @Override
    public void mousePressed(MouseEvent e) {
    }

    @Override
    public void mouseReleased(MouseEvent e) {
    }

    @Override
    public void mouseEntered(MouseEvent e) {
    }

    @Override
    public void mouseExited(MouseEvent e) {
    }
}
```

测试代码如下:

```java
public class Main {
    public static void main(String[ ] args){
        final int SIZE = 8;
        Screen screen = new Screen(SIZE);
        Dog husky = new HuskyDog(6,2);
        screen.add(husky);
        MouseListener l = new MouseAttack (husky);
        screen.addMouseListener(l);
    }
}
```

运行程序,单击鼠标左键,"哈士奇"释放攻击技能。

还可以设计许多类实现鼠标监听 MouseListener 接口,完成各种鼠标操作。例如,设计 MouseMove 类,实现单击鼠标左键,狗向上运动;单击鼠标右键,狗向下运动,代码如下:

```java
public class MouseMove implements MouseListener{
    Dog dog;
    public MouseMove(Dog dog) {
```

```java
        this.dog = dog;
    }

    @Override
    public void mouseClicked(MouseEvent e) {
        if(e.getButton() == MouseEvent.BUTTON1)          //单击鼠标左键,狗向上运动
            dog.moveUp();
        if(e.getButton() == MouseEvent.BUTTON3)          //单击鼠标右键,狗向下运动
            dog.moveDown();
    }

    @Override
    public void mousePressed(MouseEvent e) {
    }

    @Override
    public void mouseReleased(MouseEvent e) {
    }

    @Override
    public void mouseEntered(MouseEvent e) {
    }

    @Override
    public void mouseExited(MouseEvent e) {
    }
}
```

测试代码如下:

```java
public class Main {
    public static void main(String[ ] args){
        final int SIZE = 8;
        Screen screen = new Screen(SIZE);
        Dog husky = new HuskyDog(6,2);
        screen.add(husky);
        MouseListener l = new MouseMove(husky);
        screen.addMouseListener(l);
    }
}
```

运行程序,单击鼠标左键,"哈士奇"向上运动;单击鼠标右键"哈士奇"向下运动。

提示:封装、继承和多态是面向对象三大特性,是面向对象的核心,也是学习的重点,需要不断通过实践练习才能真正将其深刻理解。

6.4 内部类

Java语言允许在一个类的内部定义另外一个类,这样的类被称为内部类,内部类所在的类称为外部类,例如:

```
Class Plane{
    Class Bullet{
    }
}
```

在 Plane 飞机类中包含 Bullet 子弹类，Plane 类就是外部类，Bullet 类就是内部类。使用内部类的好处就是增强了两个类之间的联系，提高了代码的可读性和可维护性。内部类主要有静态内部类和非静态内部类两种类型，静态内部类用 static 关键字声明，非静态内部类包括成员内部类、局部内部类和匿名内部类。

6.4.1 静态内部类

静态内部类，就是使用 static 关键字修饰的成员内部类（见 6.4.2 节），静态内部类只能访问外部类的静态成员，可以不通过外部类的实例就创建一个对象。静态内部类和成员内部类的区别通过一个例子可知：

```
class Outer {
    class Inner {}
    static class StaticInner {}
}

Outer outer = new Outer();
Outer.Inner inner = outer.new Inner();
Outer.StaticInner inner0 = new Outer.StaticInner();
```

Outer 类为外部类，Inner 类为成员内部类，StaticInner 类为静态内部类。Inner 类的对象必须通过外部类对象引用才能创建，而 StaticInner 类则不需要。许多 JDK 源码中用到了静态内部类，如下一章的 HashMap 类。另外，经典的设计模式、建造者模式中也常常用到静态内部类。

【例 6.6】 设计"推箱子"游戏中的地图编辑工具。

推箱子游戏的玩法是通过"哈士奇"将"箱子"推到"圆球"所在的位置，而"树"就是障碍，会阻挡"箱子"或者"哈士奇"的运动。游戏有许多关卡，当顺利通过一关，就会进入下一关，每一关最重要的就是地图，不同的关卡有着不同的地图，如图 6.2 所示。

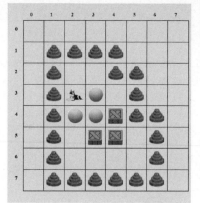
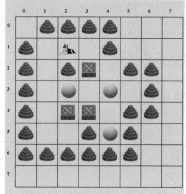

图 6.2　关卡地图

本案例的核心任务就是设计一个"地图编辑器",能够方便地制作各种有趣的地图,如图 6.3 所示。

将地图编辑器分解成箱子、障碍(树)、目标(圆球)、人物(哈士奇)等组成部分。"箱子"为 Box 类、"哈士奇"为 HuskyDog 类,在之前的例题中已经设计好。接下来完成"障碍物"类和"目标"类的设计,"障碍物"类不能移动,只需要知道它在屏幕上的位置即可,所以"障碍物"类的代码如下:

```
public class Block extends GameObject{
    public Block(int row, int col) {
        super(row,col);
    }
}
```

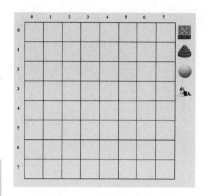

图 6.3　地图编辑器

"目标"类也一样不能移动,只需要知道在屏幕上的位置即可,代码如下:

```
public class Aim extends GameObject{
    public Aim(int row, int col) {
        super(row,col);
    }
}
```

设计完上述类之后,可以将右边的工具栏,显示在屏幕上。为了让程序更加清晰,可以设计一个工具 Toolbar 类,代码如下:

```
public class Toolbar {
    private Box box;
    private Block block;
    private Aim aim;
    private Dog dog;

    public Toolbar(Builder builder) {
        this.box = builder.box;
        this.block = builder.block;
        this.aim = builder.aim;
        this.dog = builder.dog;
    }

    public void addToScreen(Screen screen) {      // 将各个组成元素添加到屏幕中
        screen.add(box);
        screen.add(block);
        screen.add(aim);
        screen.add(dog);
    }

    public static class Builder {
        private Box box;
        private Block block;
        private Aim aim;
        private Dog dog;
```

```java
        public Builder setBox(Box box) {
            this.box = box;
            return this;
        }
        public Builder setBlock(Block block) {
            this.block = block;
            return this;
        }
        public Builder setAim(Aim aim) {
            this.aim = aim;
            return this;
        }
        public Builder setDog(Dog dog) {
            this.dog = dog;
            return this;
        }

        public Toolbar build(){
            return new Toolbar(this);
        }
    }
}
```

测试代码如下:

```java
public class Main {
    public static void main(String[] args) {
        final int SIZE = 8;
        Screen screen = new Screen(SIZE);

        Toolbar.Builder builder = new Toolbar.Builder();
        Box box = new Box(0,8);
        Block block = new Block(1,8);
        Aim aim = new Aim(2,8);
        HuskyDog husky = new HuskyDog(3,8);
        builder.setBox(box).setBlock(block).setAim(aim).setDog(husky);
        Toolbar toolbar = builder.build();
        toolbar.addToScreen(screen);
    }
}
```

运行程序,出现如图 6.3 所示的工具栏。如果想设计成其他图形的工具栏也非常方便,测试代码修改如下:

```java
public class Main {
    public static void main(String[] args) {
        final int size = 8;
        Screen screen = new Screen(size);

        Toolbar.Builder builder = new Toolbar.Builder();
        Box box = new Box(0,8);
        Block block = new Block(1,8);
```

```
        Aim aim = new Aim(2,8);
        TeddyDog teddy = new TeddyDog(3,8);
        builder.setBox(box).setBlock(block).setAim(aim).setDog(teddy);
        Toolbar toolbar = builder.build();
        toolbar.addToScreen(screen);
    }
}
```

运行程序,工具栏中的"哈士奇"就变成了"泰迪"。当需要构建的产品具备复杂创建过程时,通过建造者模式可以抽取出共性构建过程,然后交由具体实现类自定义构建流程,这样使得同样的构建行为生产出不同的产品。建造者模式分离了构建与表示,大大增加了构建产品的灵活性。

6.4.2 非静态内部类

1. 成员内部类

在类中除了可以定义成员变量和方法,还可以定义成员内部类。成员内部类可以看作外部类的一个成员,内部类和外部类可以相互访问各自所有的成员,包括私有成员。

接下来,通过一个案例来学习成员内部类的使用。

【例 6.7】 编写"飞机大战"程序,实现按键"k"控制飞机发射子弹。

将 Bullet 子弹类设计成 Plane 类的内部类,代码如下:

视频讲解

```
public class Plane extends GameObject implements Move{
    public Plane(int row, int col) {
        super(row,col);
    }

    @Override
    public void moveUp() {
        setRow(getRow() - 1);
    }

    @Override
    public void moveDown() {
        setRow(getRow() + 1);
    }

    @Override
    public void moveLeft() {
        setCol(getCol() - 1);
    }

    @Override
    public void moveRight() {
        setCol(getCol() + 1);
    }

    public Bullet shoot() {
        Bullet bullet = new Bullet(getRow(),getCol());
        return bullet;
    }
```

```java
public class Bullet extends GameObject {
    public Bullet(int row, int col) {
        super(row,col);
    }
    public void moveUp() {
        setRow(getRow() - 1);
    }
}
```

测试代码如下:

```java
public class Main {
    public static void main(String[ ] args){
        final int size = 8;
        Screen screen = new Screen(size);
        Plane plane = new Plane(7,4);
        Plane.Bullet bullet = null;
        screen.add(plane);
        
        char key;
        while(true) {
            screen.delay();
            key = screen.getKey();
            
            if(key == 'w') {
                plane.moveUp();
            }
            if(key == 's') {
                plane.moveDown();
            }
            if(key == 'a') {
                plane.moveLeft();
            }
            
            if(key == 'd') {
                plane.moveRight();
            }
            
            if(key == 'k') {
                if(bullet == null) {
                    bullet = plane.shoot();
                    screen.add(bullet);
                }
            }
            
            if(bullet != null) {
                bullet.moveUp();
            }
        }
    }
}
```

运行程序,按下按键"k",飞机发射子弹。

2. 局部内部类

局部内部类指的是在方法体内或语句块内定义的类,它和局部变量一样,其作用只限于方法内部。局部内部类中的变量和方法只能在创建该局部内部类的方法中访问,不能有任何访问修饰符,局部内部类的基本格式如下:

```
修饰符 class 外部类名称{
    修饰符 返回值类型 外部类方法名([参数列表]){
        class 局部内部类名称{
        }
    }
}
```

在实际开发中,局部内部类使用较少,只需了解即可。

3. 匿名内部类

如果某个类只创建一个对象时,可以使用匿名内部类。匿名内部类就是没有名字的内部类。在调用包含有接口类型参数的方法时,为了简便,通常直接通过匿名类的形式传入接口类型的参数,在匿名类中直接完成方法的实现。创建匿名内部类的基本格式如下:

```
new 类名或接口名(){
    //重写方法
}
```

匿名内部类本质就是继承该类或者实现接口的子类匿名对象。例6.5中鼠标控制"哈士奇"释放攻击技能,利用匿名类实现的代码如下:

```java
public class Main {
    public static void main(String[ ] args){
        final int size = 8;
        Screen screen = new Screen(size);
        Dog husky = new HuskyDog(6,2);
        screen.add(husky);

        screen.addMouseListener(new MouseListener (){
            public void mouseClicked(MouseEvent e) {
                husky.attack();
            }

            @Override
            public void mousePressed(MouseEvent e) {
            }

            @Override
            public void mouseReleased(MouseEvent e) {
            }

            @Override
            public void mouseEntered(MouseEvent e) {
            }

            @Override
```

```
            public void mouseExited(MouseEvent e) {
            }
        });
    }
}
```

运行程序,鼠标控制"哈士奇"释放攻击技能。对比例6.5的代码可知,只创建一个类对象时,使用匿名内部类可以使程序简洁。

当一个接口有多个抽象方法的时候,每次使用匿名内部类仍然需要将每个抽象方法重写,非常不方便。Java 8中引入了一个重要的新特性Lambda表达式,可以取代大部分的匿名内部类代码,使其变得更简洁紧凑。Lambda表达式又称为匿名方法或者闭包,简而言之就是匿名的方法,语法格式如下:

```
(参数列表)->{表达式主体}
```

例如,计算两个整数的方法,代码如下:

```
public void getSum(int x, int y){
    return x + y;
}
```

写成Lambda表达式,代码如下:

```
(x,y) -> x + y;
```

使用Lambda表达式可以简化代码,将多行代码压缩成一行。

使用匿名类完成例6-3。使用Lambda表达式实现步骤:先定义一个接口,然后将需要使用的方法删掉,其余方法使用default修饰重写,这样就能使用Lamdba表达式简化代码。

接口代码如下:

```
interface ClickedListener extends MouseListener{
    @Override
    public default void mouseEntered(MouseEvent e) {}

    @Override
    public default void mouseExited(MouseEvent e) {}

    @Override
    public default void mousePressed(MouseEvent e) {}

    @Override
    public default void mouseReleased(MouseEvent e) {}
}
```

mouseClicked()方法没有被重写,所以可以使用Lamdba表达式使用该方法,代码如下:

```
public class Main {
    public static void main(String[] args){
        final int size = 8;
        Screen screen = new Screen(size);
        Dog husky = new HuskyDog(6,2);
```

```
        screen.add(husky);
        MouseListener l = new MouseMove(husky);
        screen.addMouseListener((ClickedListener)(e) -> husky.attack());
    }
}
```

运行程序,单击鼠标可以控制"哈士奇"发动攻击技能。对比匿名内部类和 Lamdba 表达式实现接口方法,明显看到 Lambda 表达式更加简洁和清晰。

6.5 综合案例:"地图"编辑器

视频讲解

推箱子是一个经典的益智游戏,可以训练使用者的逻辑思考能力。在一个狭小的仓库中,要求把木箱推到指定的位置,稍不小心就会出现箱子无法移动或者通道被堵住的情况,所以需要巧妙地利用有限的空间和通道,合理安排移动的次序和位置,才能顺利地完成任务。

在例 6.6 中已经实现了工具栏的设置,接下来需要完成鼠标拖拽实现编辑地图的工具。首先需要修改例 6.6 中的 Toolbar 类,新增 getSelect()方法,实现根据鼠标单击的位置获得选中的物体,代码如下:

```java
public class Toolbar {
    private Box box;
    private Block block;
    private Aim aim;
    private Dog dog;

    public Toolbar(Builder builder) {
        this.box = builder.box;
        this.block = builder.block;
        this.aim = builder.aim;
        this.dog = builder.dog;

    }

    public void addToScreen(Screen screen) {
        screen.add(box);
        screen.add(block);
        screen.add(aim);
        screen.add(dog);
    }

    public GameObject getSelect(int row, int col) {           //获得鼠标选中的元素
        GameObject object = null;
        if(box.getRow() == row && box.getCol() == col) {
            object = new Box(row,col);
        }
        if(aim.getRow() == row && aim.getCol() == col) {
            object = new Aim(row,col);
        }
        if(block.getRow() == row && block.getCol() == col) {
            object = new Block(row,col);
        }
```

```java
            if(dog.getRow() == row && dog.getCol() == col) {
                object = new HuskyDog(row,col);
            }
            return object;
        }

        public static class Builder {
            private Box box;
            private Block block;
            private Aim aim;
            private Dog dog;

            public Builder setBox(Box box) {
                this.box = box;
                return this;
            }
            public Builder setBlock(Block block) {
                this.block = block;
                return this;
            }
            public Builder setAim(Aim aim) {
                this.aim = aim;
                return this;
            }
            public Builder setDog(Dog dog) {
                this.dog = dog;
                return this;
            }

            public Toolbar build(){
                return new Toolbar(this);
            }
        }
    }
```

然后,设计鼠标类实现鼠标接口,鼠标操作的核心内容是选择合适的对象,并将其放置在合适的位置。因此,需要重写鼠标被按下和释放的代码,根据鼠标按下的位置可以获得选中的对象,然后根据鼠标释放的位置可以获得放置对象的位置,最后将新生成的对象添加到屏幕上。代码如下:

```java
public class Main {
    public static void main(String[ ] args){
        final int SIZE = 8;
        Screen screen = new Screen(SIZE);
        Toolbar.Builder builder = new Toolbar.Builder();
        Box box = new Box(0,8);
        Block block = new Block(1,8);
        Aim aim = new Aim(2,8);
        HuskyDog husky = new HuskyDog(3,8);
        builder.setBox(box).setBlock(block).setAim(aim).setDog(husky);
        Toolbar toolbar = builder.build();
        toolbar.addToScreen(screen);
```

```java
screen.addMouseListener(new MouseListener() {
    GameObject object = null;

    @Override
    public void mouseClicked(MouseEvent e) {
    }

    @Override
    public void mousePressed(MouseEvent e) {                // 获得选中的对象
        int row = screen.getRow(e.getY());
        int col = screen.getCol(e.getX());
        object = toolbar.getSelect(row, col);
    }

    @Override
    public void mouseReleased(MouseEvent e) {               //获得选中的位置
        int row = screen.getRow(e.getY());
        int col = screen.getCol(e.getX());
        if(object != null) {
            object.setRow(row);
            object.setCol(col);
            screen.add(object);
        }
    }

    @Override
    public void mouseEntered(MouseEvent e) {
    }

    @Override
    public void mouseExited(MouseEvent e) {
    }
});
}
}
```

运行程序，按下鼠标选中合适的对象拖到"屏幕"中，释放鼠标，对象就被添加到"屏幕"上的相应位置。

完成了任务，可以进一步优化代码。如果不小心将对象放错了位置，可以再次选中该对象，然后拖到"屏幕"外，就从"屏幕"中删掉。实现该功能只需要修改匿名内部类即可，代码如下：

```java
public class Main {
    public static void main(String[ ] args){
        final int SIZE = 8;
        Screen screen = new Screen(SIZE);
        Toolbar.Builder builder = new Toolbar.Builder();
        Box box = new Box(0,8);
        Block block = new Block(1,8);
        Aim aim = new Aim(2,8);
        HuskyDog husky = new HuskyDog(3,8);
        builder.setBox(box).setBlock(block).setAim(aim).setDog(husky);
        Toolbar toolbar = builder.build();
```

```java
            toolbar.addToScreen(screen);

            screen.addMouseListener(new MouseListener() {
                GameObject object = null;

                @Override
                public void mouseClicked(MouseEvent e) {
                }

                @Override
                public void mousePressed(MouseEvent e) {
                    int row = screen.getRow(e.getY());
                    int col = screen.getCol(e.getX());
                    object = toolbar.getSelect(row, col);
                    if(object == null && screen.isInside(row, col)) {
                        object = screen.getObject(row, col);
                    }
                }

                @Override
                public void mouseReleased(MouseEvent e) {
                    int row = screen.getRow(e.getY());
                    int col = screen.getCol(e.getX());
                    if(object != null) {
                        if(screen.isInside(row, col)) {                //如果鼠标在屏幕内
                            object.setRow(row);
                            object.setCol(col);
                            screen.add(object);
                        }
                        else {
                            screen.remove(object);
                        }
                    }
                }

                @Override
                public void mouseEntered(MouseEvent e) {
                }

                @Override
                public void mouseExited(MouseEvent e) {
                }
            });
        }
    }
```

运行程序,可以通过鼠标轻松地设计各种有趣的地图。

根据鼠标选中的位置信息生成不同的对象,也可以使用设计模式中的工厂模式产生对象,这样使代码具有更好的扩展性,有兴趣的读者可以自己尝试一下。另外,设计的地图如何长期保存?这些内容将会在后续的章节陆续讲解。

习题

6.1 以下 Java 语言中关于继承的叙述正确的是_____。
　　A. 类只允许单一继承
　　B. 一个类只能实现一个接口
　　C. 一个类不能同时继承一个类和实现一个接口
　　D. 接口只允许单一继承

6.2 类中的成员方法被以下_____修饰符修饰,该方法只能在本类被访问?
　　A. public　　　　B. default　　　　C. protected　　　　D. private

6.3 在使用 super 和 this 关键字时,以下描述正确的是_____。
　　A. 在子类构造方法中使用 super()显式调用父类的构造方法,super()必须写在子类构造方法的第一行,否则编译不通过
　　B. super()和 this()不一定要放在构造方法内第一行
　　C. this()和 super()可以同时出现在一个构造函数中
　　D. this()和 super()可以在 static 环境中使用,包括 static 方法和 static 语句块

6.4 以下对抽象类的描述正确的是_____。
　　A. 抽象类没有构造方法
　　B. 抽象类必须提供抽象方法
　　C. 有抽象方法的类一定是抽象类
　　D. 抽象类可以通过 new 关键字直接实例化

6.5 以下对接口的描述错误的有_____。
　　A. 接口没有提供构造方法
　　B. 接口中的方法默认使用 public abstract 修饰
　　C. 接口中的属性默认使用 public static final 修饰
　　D. 接口不允许多重继承

6.6 下列选项中关于 Java 中 super 关键字的说法错误的是_____。
　　A. super 关键字是在子类对象内部指代其父类对象的引用
　　B. super 关键字不仅可以指代子类的直接父类,还可以指代父类的父类
　　C. 子类可以通过 super 关键字调用父类的方法
　　D. 子类可以通过 super 关键字调用父类的属性

6.7 设计程序,实现单击鼠标控制"哈士奇"移动到鼠标所在的位置。

第二部分

提高篇

第7章

集合与泛型

在编写程序时，经常要用到一组同一类型的对象，使用数组可以保存多个同一类型的对象，但是数组一经定义之后，长度便不能改变。面对数据不确定的情况下，数组无法灵活应对。例如，在一个公司的员工管理系统中，员工数目是动态变化的，随时都可能有新招或者离职的员工。为了应对这种数据动态变化的情况，Java 语言提供了一套接口和类构成的集合框架，可以方便地处理多个对象。集合可以存储任何类型的对象，并且长度可变，相比于数组，集合更加灵活。

7.1 集合的概念

集合又被称为容器，可以存放不同类型、不限数量的对象，集合能够帮助程序设计者管理对象。Java 语言中的集合类按照存储结构可以分为两大类：实现 Collection 接口的类和实现 Map 接口的类。

Collection 接口是单列集合的根接口，每次存储元素都是单列对象。Collection 集合有两个重要的子接口，分别是 List(列表)和 Set(集)。List 接口的特点是元素有序且可重复；Set 接口的特点正相反，元素无序且不可重复。List 接口常用实现类有 ArrayList 类和 LinkedList 类；Set 接口主要实现类有 HashSet 类和 TreeSet 类。

Map 接口是双列集合的根接口，每次存储的元素都包含成对的键(Key)-值(Value)对象。Key 是唯一的，使用 Map 集合时可以用指定的 Key 获取对应的 Value。Map 接口的主要实现类有 HashMap 类和 TreeMap 类。

集合的框架体系如图 7.1 所示。

图 7.1 集合框架体系图

7.2 Collection 接口与实现类

Collection 接口是单列集合的根接口,统一定义了单列集合的常用方法,主要方法及作用如表 7.1 所示。

表 7.1 Collection 接口的主要方法及作用

方　　法	作　　用
boolean add(Object element)	增加元素到集合中
boolean remove(Object o)	删除集合中的某个元素
void clear()	清空集合
int size()	获取集合中元素的个数
boolean isEmpty()	判断该集合是否为空
Iterator iterator()	获得在此 collection 的元素上进行迭代的迭代器
boolean contains(Object element)	判断集合是否包含某个元素

对于初学者,掌握常用的方法即可。例如,增加元素到容器中,从容器中删除元素,判断容器是否为空,获得元素的数量,获得迭代器用于遍历所有元素等。

7.2.1 List 接口与实现类

List 接口是 Collection 接口的子接口,实现了一种线性表的数据结构。List 接口的特点是元素有序且可重复。List 接口可以对列表中每个元素的插入位置进行精确控制,能够像数组一样通过下标访问其中的元素。List 接口除了继承 Collection 接口的方法外,还定义了自己的方法,如表 7.2 所示。

表 7.2 List 接口的主要方法及作用

方　　法	作　　用
void add(int index,Object element)	在 index 位置插入元素
boolean addAll(int index,Collection elements)	从 index 位置开始将集合中的所有元素添加进来
Object get(int index)	获取指定 index 位置的元素
int indexOf(Object element)	返回元素在集合中首次出现的位置
int lastIndexOf(Object element)	返回元素在当前集合中末次出现的位置

续表

方　　法	作　　用
Object remove(int index)	移除指定 index 位置的元素,并返回此元素
Object set(int index,Object element)	设置指定 index 位置的元素为 element,并返回此元素
List subList(int fromIndex,int toIndex)	返回从 fromIndex 到 toIndex 位置的子集合

List 接口的实现类都可以调用上述方法对集合中的元素进行操作。ArrayList 类和 LinkedList 类都是 List 接口的常用实现类,都是程序中常用的集合。

1. ArrayList 类

ArrayList 类是最常用的 List 接口实现类,本质上它就是一个长度可变的数组,其元素可以动态地增加和删除。它的优点是可以根据索引位置对集合进行快速的随机访问,缺点是向指定的索引位置插入对象或删除对象的速度较慢。

【例 7.1】 游戏中有许多英雄,随着版本不断升级,游戏中会出现新的英雄,编写程序设计英雄库。

随着版本升级,英雄库里会出现新的英雄,所以使用 List 集合方便存储各种英雄信息。例如,假设英雄库里已经有了"哈士奇""泰迪"等英雄,新增一种"柯基"英雄,新增柯基类代码如下:

视频讲解

```
public class CorgiDog extends Dog{
    public CorgiDog(int row, int col) {
        super(row, col);
    }
}
```

测试代码如下:

```
public class Main {
    public static void main(String[ ] args){
        final int SIZE = 8;
        Screen screen = new Screen(SIZE);

        ArrayList heros = new ArrayList();
        HuskyDog husky = new HuskyDog(0, 7);
        TeddyDog teddy = new TeddyDog(1, 7);
        CorgiDog corgi = new CorgiDog(2, 7);
        heros.add(husky);
        heros.add(teddy);
        heros.add(corgi);
        screen.add(heros);

        screen.delay();
        Dog dog = (Dog) heros.get(2);
        dog.moveLeft();
    }
}
```

运行程序,屏幕上出现三种"英雄",并且"柯基"向左运动一格。使用 List 集合可以非常容易实现动态地增加新元素,并且通过索引获得对应位置的元素。

上述程序虽然能正常运行,但是编写的时候,在代码的左边会出现带感叹号的提示信息。例如,在"ArrayList heros= new ArrayList();"语句旁边就出现了带感叹号的警告信息,将鼠标移到感叹号上出现提示信息:"ArrayList 是原始类型。应该将对通用类型 ArrayList<E> 的引用参数化"。这是因为使用 ArrayList 集合时并没有显式地指定集合中存储元素的类型,这会产生安全隐患,涉及泛型安全机制的问题,将会在后面的章节详细介绍,暂时可以忽略该问题。

2. LinkedList 类

ArrayList 集合的存储结构是数组,在内部增删元素效率非常低。如果需要经常在内部添加或者删除元素,可以使用 LinkedList 集合。该集合的存储结构是一个双向循环链表,增加或者删除元素的效率非常高。LinkedList 集合与 ArrayList 集合相比,前者增加或者删除特定位置元素的效率非常高,后者查找元素的效率非常高。

LinkedList 集合除了继承 Collection 和 List 接口的方法外,还定义了许多首尾操作的方法,如表 7.3 所示。

表 7.3 LinkedList 集合主要方法及作用

方 法	作 用
public void addFirst(E element)	将指定的元素插入此列表的开头
public void addLast(E element)	将指定元素添加到此列表的结尾
public void push(E element)	将元素推入此列表所表示的堆栈
public E getFirst()	返回此列表的第一个元素
public E getLast()	返回此列表的最后一个元素
public E removeFirst()	移除并返回此列表的第一个元素
public E removeLast()	移除并返回此列表的最后一个元素
public E pop()	从此列表所表示的堆栈处弹出一个元素

下面通过一个例子演示 LinkedList 的使用。

视频讲解

【例 7.2】 编写程序,实现按键控制"贪吃蛇"上、下、左、右运动,如图 7.2 所示。

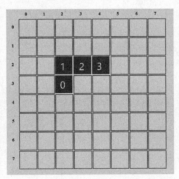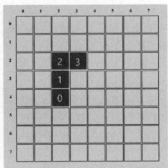

图 7.2 "贪吃蛇"运动轨迹示意图

"贪吃蛇"吃到"食物"后身体会变长,使用 List 集合保存"贪吃蛇"的数据,会非常容易实现。仔细观察图 7.2,对比运动前后的图像会发现,只有一个方块的位置发生了改变,也就是"蛇身"最后一个方块,它仿佛直接挪到"蛇头"将运动到的地方。用数组很难实现这样的过程,但是使用 List 集合就很容易,相当于向 List 集合中新增一个元素,并删除掉最后一个元素,中间的元素都不需要变化,代码如下:

```java
public class Snake {
    private LinkedList cells;
    private int dir;
    
    public Snake() {
        cells = new LinkedList();
        dir = 2;
        cells.add(new Cell(0,4));
        cells.add(new Cell(0,5));
        cells.add(new Cell(0,6));
    }
    
    public void move() {
        Cell cell = new Cell((Cell) cells.getFirst());    //生成一个新的节点作为蛇头
        if(dir == 1) {
            cell.moveUp();
        }
        if(dir == 2) {
            cell.moveDown();
        }
        if(dir == 3) {
            cell.moveLeft();
        }
        if(dir == 4) {
            cell.moveRight();
        }
        cells.addFirst(cell);         //将新的蛇头添加到列表中
        cells.removeLast();           //移出列表中最后一个元素
    }
    
    public LinkedList getCell() {
        return cells;
    }
    
    public void setDirection(int d) {
        dir = d;
    }
}
```

测试代码如下：

```java
public class Main {
    public static void main(String[] args){
        Screen screen = new Screen(8);
        Snake snake = new Snake();
        screen.add(snake.getCell());
        
        char key;
        while(true) {
            screen.delay();
            key = screen.getKey();
            if(key == 'w') {
```

```
                snake.setDirection(1);
            }
            if(key == 's') {
                snake.setDirection(2);
            }
            if(key == 'a') {
                snake.setDirection(3);
            }
            if(key == 'd') {
                snake.setDirection(4);
            }
            screen.remove(snake.getCell());
            snake.move();
            screen.add(snake.getCell());
        }
    }
}
```

通过 LinkedList 集合,使用 addFirst()、removeLast()等方法对集合中的元素进行增删,从而实现"贪吃蛇"的运动。由此可见,使用 LinkedList 集合对元素进行增删操作非常方便。

7.2.2 Set 接口与实现类

Set 接口与 List 接口一样,也是 Collection 接口的子接口。Set 接口并没有对 Collection 接口进行功能上的扩充,只是比 Collection 接口更加严格,Set 接口中的元素不能出现重复。Set 接口的主要实现类有 HashSet 类和 TreeSet 类。其中,HashSet 类是根据对象的哈希值来确定元素在集合中的存储位置,因此存储或者查找元素的效率非常高。TreeSet 类则是以二叉树的方法存储元素,它可以实现对集合中的元素进行排序操作。

1. HashSet 类

HashSet 类是 Set 接口的一个实现类,它所存储的元素是不可重复的。向 HashSet 类中新增元素时,首先调用当前新增元素的 hashCode()方法获得对象的哈希值,然后根据对象的哈希值计算出一个存储位置。如果该位置上没有元素,则直接存入元素;如果该位置上有元素,则会调用 equals()方法让当前存入的元素依次和该位上的元素进行比较,如果比较的结果为"false",则将该元素存入集合,否则将元素舍弃。为了保证存入 HashSet 类的元素不重复,存入对象的时候,需要重写 Object 类的 hashCode()和 equals()方法。

例如,"贪吃蛇"游戏中,如果增加多个"食物",并且要保证每个"食物"的位置都不相同,这样可以避免"贪吃蛇"在一个位置吃到"食物"后"食物"没有消失的情况。为了保证所有的"食物"位置都不相同,可以使用 HashSet 集合保存"食物"数据。

视频讲解

【例 7.3】 编写程序,实现"贪吃蛇"游戏中多个"食物"随机出现在屏幕上,并且"食物"的位置都不相同。

为了简化处理,先增加两个位置相同的"食物",代码如下:

```java
public class Main {
    public static void main(String[ ] args){
        HashSet set = new HashSet();
        Screen screen = new Screen(8);
        Food food1 = new Food(2,3);
        Food food2 = new Food(2,3);
        set.add(food1);
        set.add(food2);
        screen.add(set);
    }
        screen.delay();
        screen.remove(food1);
    }
}
```

运行程序,屏幕上出现的"食物"并没有消失。在代码中新增一句"screen.remove(food2);",再运行程序,则"食物"出现在屏幕上,然后消失。上述代码意味着虽然两个"食物"位置相同,但是都成功添加到了集合中。为了解决这一问题,需要对 Food 类进行修改,重写 hashCode() 和 equals() 方法,对于位置相同的"食物"就可以判断为同一个"食物",修改后的代码如下:

```java
public class Food extends GameObject{
    public Food(int r, int c) {
        super(r,c);
    }

    public int hashCode() {
        return this.getRow() + this.getCol();
    }

    public boolean equals(Object obj) {
        if(!(obj instanceof Food)) {
            return false;
        }
        Food food = (Food) obj;
        return this.getRow() == food.getRow() && this.getCol() == food.getCol();
    }
}
```

再次运行程序,虽然增加了两个"食物",但是只删除一次,"食物"就消失了,也就意味着位置相同的"食物"不会重复增加进 HashSet 类之中。

修改后的 Food 类重写了 Object 类的 hashCode() 和 equals() 方法。在 equals() 方法里比较对象的位置是否相等,并返回结果。当 HashSet 集合新增元素的时候,发现哈希值相同,并且调用 equals() 方法比较位置信息也相同,则 HashSet 集合认为两个对象相同,因此重复的 Food 对象被舍弃了。

2. TreeSet 类

TreeSet 类是 Set 接口的另一个实现类,内部采用平衡二叉树存储元素,这样可以保证

TreeSet 类中既没有重复的元素,而且可以对元素进行排序。与 HashSet 类相比,TreeSet 类增加了一些方法,如表 7.4 所示。

表 7.4 TreeSet 类的主要方法及作用

方 法	作 用
Comparator comparator():	如果 TreeSet 采用了定制顺序,则该方法返回定制排序所使用的 Comparator;如果 TreeSet 采用自然排序,则返回 null
Object first():	返回集合中的第一个元素
Object last():	返回集合中的最后一个元素
Object lower(Object element):	返回指定元素之前的元素
Object higher(Object element):	返回指定元素之后的元素

7.2.3 Collection 集合遍历

在程序开发过程中,对集合进行遍历是常见的操作。遍历集合中元素的方法有两种:使用 Iterator 迭代器对象和使用增强 for 循环。

1. 使用迭代器

迭代器是一个可以遍历集合中每个元素的对象,主要用于迭代访问。迭代器能提供一种访问集合对象中每个元素的途径,同时又不需要暴露该对象的内部细节。Java 语言提供 Iterator 迭代器和 Iterable 接口来实现集合类的可迭代性,主要方法如表 7.5 所示。

表 7.5 Iterator 的主要方法及作用

方 法	作 用
boolean hasNext()	检查集合中是否还有元素
E next()	返回迭代器的下一个元素,并且更新迭代器的状态
void remove()	将迭代器返回的元素删除

视频讲解

【例 7.4】 编写程序,实现"贪吃蛇"游戏中的"食物"随机出现在屏幕上,但不能出现在"贪吃蛇"所在位置。

要实现"食物"的位置不能与"贪吃蛇"的位置重叠,就需要遍历"贪吃蛇"的每一个方块,确保每一个方块的位置都不与"食物"位置相同。集合的遍历可以通过迭代器对象实现,而迭代器对象可以通过调用集合对象的 iterator()方法获得。

"食物类"代码如下:

```
public class Food extends GameObject{
    public Food(int row, int col) {
        super(row,col);
    }
}
```

"食物生产工厂类"代码如下:

```
public class Factory {
    private LinkedList cells;

    public boolean isSame(int row, int col) {                              //判断位置是否重复
```

```java
            Iterator iterator = cells.iterator();
            boolean isMul = false;
            
            while(iterator.hasNext()) {                     //遍历每一个元素
                Cell cell = (Cell) iterator.next();
                if(cell.getRow() == row && cell.getCol() == col) {        //位置相同时
                    isMul = true;
                    break;
                }
            }
            return isMul;
        }
        
        public Food getFood(LinkedList cells) {
            this.cells = cells;
            Random rand = new Random();
            int row = rand.nextInt(Config.SCREENSIZE);
            int col = rand.nextInt(Config.SCREENSIZE);
            
            while(isSame(row,col)) {                        //如果位置重复了,则产生新的随机位置
                row = rand.nextInt(Config.SCREENSIZE);
                col = rand.nextInt(Config.SCREENSIZE);
            }
            
            return new Food(row,col);
        }
}
```

测试代码如下:

```java
public class Main {
    public static void main(String[] args){
        Factory factory = new Factory();
        Screen screen = new Screen(Config.SCREENSIZE);
        Snake snake = new Snake();
        
        screen.add(snake.getCell());
        Food food = factory.getFood(snake.getCell());
        screen.add(food);
        char key;
        while(true) {
            screen.delay();
            key = screen.getKey();
            if(key == 'w') {
                snake.setDirection(1);
            }
            if(key == 's') {
                snake.setDirection(2);
            }
            if(key == 'a') {
                snake.setDirection(3);
            }
```

```
            if(key == 'd') {
                snake.setDirection(4);
            }
            
            screen.remove(snake.getCell());
            snake.move();
            screen.add(snake.getCell());
        }
    }
}
```

2. 使用增强 for 循环

虽然迭代器可以用来遍历集合中的元素,但是写法比较烦琐。Java 语言提供了 foreach 循环遍历集合,它是一种更为简洁的 for 循环,也称增强 for 循环,语法格式如下:

```
for(容器中元素类型 临时变量:容器变量){
    //语句;
}
```

如果使用 foreach 循环可将 Factory 类中的 isSame()方法修改为:

```
public boolean isSame(int row, int col) {
    boolean isMul = false;
    for(Object object:cells) {
        Cell cell = (Cell )object;
        if(cell.getRow() == row && cell.getCol() == col) {
            isMul = true;
            break;
        }
    }
    return isMul;
}
```

相比使用迭代器对象遍历集合,foreach 循环更为简洁,无须循环条件,也没有迭代语句。

7.3 Map 接口与实现类

Map 集合使用键值对来存储数据,将键映射到值对象,一个映射不能包含重复的键,每一个键最多只能映射到一个值,这样在访问 Map 集合中的元素时,只要指定了键,就能找到对应的值。Map 接口的常用方法如表 7.6 所示。

表 7.6　Map 接口的常用方法及作用

方　　法	作　　用
Object put(Object key,Object value)	向集合中添加一个键值对
Object get(Object key)	返回指定键的值
Object remove(Object key)	从集合中删除指定键的键值对

续表

方法	作用
boolean containsKey(Object key)	判断集合中是否包含指定的键
boolean containsValue(Object value)	判断集合中是否包含指定的值

Map 接口常用的实现类为 HashMap 类和 TreeMap 类。HashMap 类是基于哈希表实现，对于元素的增、删、改、查等操作效率较高。TreeMap 类是基于红黑树实现，其中的元素是按照某种顺序排列，对于元素的增、删、改、查等操作的效率没有 HashMap 类高。如果没有特别的要求，通常情况下使用的是 HashMap 类，在第 13 章网络程序中多台计算机之间相互通信就是通过 HashMap 类实现。

7.4 泛型

集合可以存储任意类型的对象元素，当一个对象存入集合后，从集合中取出该对象时，这个对象的编译类型就统一变成了 Object 类型。这样会存在一个问题，取出元素进行强制类型转换就容易出现错误。例如，存放"食物"的 Set 集合，不小心将"方块"也存进去了，代码如下：

```java
public class Main {
    public static void main(String[ ] args){
        HashSet set = new HashSet();
        Screen screen = new Screen(8);
        Food food1 = new Food(2,3);
        Food food2 = new Food(2,3);
        Cell cell = new Cell(4,5);
        set.add(food1);
        set.add(food2);
        set.add(cell);
        for(Object object:set) {
            Food obj = (Food) object;
            screen.add(obj);
        }
    }
}
```

编译的时候不会提示错误，因为集合可以存放任意类型对象，但是在运行的时候程序会出现异常，原因是不能将存储在 Set 集合中的 Cell 对象转换成 Food 对象。当无法确定一个集合中的元素到底是什么类型时，取出元素进行强制类型转换就容易出现错误。为了解决这个问题，Java 语言引入了"参数化类型"的概念，也就是泛型。泛型可以限定操作的数据类型，在定义集合类时，可以使用"<参数化类型>"的方法指定该集合中存储的数据类型，例如：

```java
Hashset<Food> set = new Hashset<Food>();
```

使用泛型限定 Hashset 集合里只能存储 Food 类型的数据。修改代码之后，程序在编译期就会提示错误。并且程序使用了泛型去规定 Hashset 集合只能存入 Food 类型元素，遍历集合元素时，可以指定元素的类型为 Food，而不是 Object，这样就避免了在程序中进行强制类型转换，代码

如下:

```java
public class Main {
    public static void main(String[ ] args){
        HashSet<Food> set = new HashSet<Food>();
        Screen screen = new Screen(8);
        Food food = new Food(2,3);
        set.add(food);
        for(Food object:set) {
            screen.add(object);
        }
    }
}
```

使用泛型可以统一数据类型,便于操作。将运行时的异常提前到了编译时出现,提高了效率。另外,还可以避免进行强制类型转换,实现代码的模板化,把数据类型当作参数传递,提高了代码的可重用性。

视频讲解

7.5 综合案例:"飞机大战"游戏

"飞机大战"是一款经典的射击操作类游戏,按键"w""s""a""d"控制飞机上、下、左、右移动,按键"K"控制飞机发射子弹。本案例与例 6.6 的区别在于,新飞机能发射无限子弹流。

新游戏中飞机类和子弹类不需要修改,代码如下:

```java
public class Plane extends GameObject implements Move{
    public Plane(int r, int c) {
        super(r,c);
    }

    @Override
    public void moveUp() {
        setRow(getRow() - 1);
    }

    @Override
    public void moveDown() {
        setRow(getRow() + 1);
    }

    @Override
    public void moveLeft() {
        setCol(getCol() - 1);
    }

    @Override
    public void moveRight() {
        setCol(getCol() + 1);
    }
```

```java
    public Bullet shoot() {
        Bullet bullet = new Bullet(getRow(),getCol());
        return bullet;
    }

    public class Bullet extends GameObject {
        public Bullet(int row, int col) {
            super(row,col);
        }
        public void moveUp() {
            setRow(getRow() - 1);
        }
    }
}
```

然后,设计"飞机大战"游戏类 PlaneGame,用集合存储子弹的信息,代码如下:

```java
public class PlaneGame extends GameCore{
    private Screen screen;
    Plane plane;
    ArrayList<Plane.Bullet> bullets = new ArrayList<Plane.Bullet>();
    Plane.Bullet bullet;

    @Override
    public void initSprite() {
        final int SIZE = 8;
        screen = new Screen(SIZE);
        plane = new Plane(7,4);
        screen.add(plane);
        bullet = null;
    }

    @Override
    public void update() {
        char key = screen.getKey();
        screen.delay();

        if(key == 'w') {
            plane.moveUp();
        }
        if(key == 's') {
            plane.moveDown();
        }
        if(key == 'a') {
            plane.moveLeft();
        }
        if(key == 'd') {
            plane.moveRight();
        }

        if(key == 'k') {
            bullet = plane.shoot();
```

```
            screen.add(bullet);
            bullets.add(bullet);
        }

        for(Plane.Bullet blet:bullets) {           // 已经发射的子弹继续向上运动
            blet.moveUp();
        }
    }
}
```

测试代码如下：

```
public class Main {
    public static void main(String[ ] args){
        PlaneGame game = new PlaneGame();
        game.init();
        game.run();
    }
}
```

使用集合保存子弹，可以实现无限子弹流，只需要按下按键"K"就能发射新的子弹。

习题

7.1 如果要求不能包含重复的元素，使用_____结构存储最合适。
   A. Collection   B. List   C. Set   D. Map

7.2 如果要求不能包含重复的元素，并且按照一定的顺序排列，使用_____结构存储最合适。
   A. LinkedList   B. ArrayList   C. hashSet   D. TreeSet

7.3 Java 中，_____接口以键-值对的方式存储对象。
   A. java.util.Collection   B. java.util.Map
   C. java.util.List   D. java.util.Set

7.4 Java 中的集合类包括 ArrayList、LinkedList、HashMap 等类，下列关于集合类描述错误的是_____。
   A. ArrayList 和 LinkedList 均实现了 List 接口
   B. ArrayList 的访问速度比 LinkedList 快
   C. 添加和删除元素时，ArrayList 的表现更佳
   D. HashMap 实现 Map 接口，它允许任何类型的键和值对象，并允许将 null 用作键或值

7.5 编写程序，利用集合实现"贪吃蛇"游戏。

第8章

异 常 处 理

程序在运行的过程中难免会遇到一些特殊情况,例如,磁盘空间不足、数组下标越界等。这些由外部原因导致程序在运行时发生特殊情况的事件被称为异常。Java 语言提供了优秀的异常机制,让程序在发生异常时,能够有效处理异常,使程序尽可能恢复正常并继续运行。

8.1 异常处理的方法

8.1.1 异常的概念

异常是程序在运行过程中发生的非正常情况,它可能使程序无法正常运行。接下来通过一个简单的例子来认识异常。

【例 8.1】 常见异常示例。

当某个对象的引用为 null 时,调用该对象的成员方法会产生异常,代码如下:

视频讲解

```
public class Main {
    public static void main(String[ ] args) {
        HuskyDog dog = null;
        dog.moveUp();
        Screen screen = new Screen(8);
        HuskyDog dog1 = new HuskyDog(4,5);
        screen.add(dog1);
    }
}
```

程序的运行结果如图 8.1 所示。

从运行结果可知,程序产生了一个"NullPointerException"异常。这个异常发生之后,程序会立即结束,无法继续向下执行。

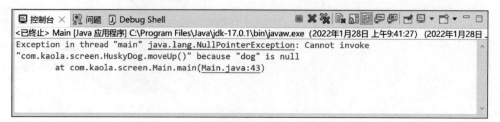

图 8.1 异常示例的运行结果

图 8.2 异常类的继承体系

除了"NullPointerException"异常外,Java语言还提供了大量的异常类,Java语言通过面向对象的方法来处理异常,为各种异常建立了对应的类。异常类的继承体系如图8.2所示。

由图 8.2 可知,异常类都是 Throwable 类的子类。Throwable 类有两个直接子类:Error 类和 Exception 类,Error 类代表程序中产生的错误,Exception 代表程序中产生的异常。

1. Error 类

Error 类为错误类,它表示程序运行时产生的系统内部错误,如系统崩溃、堆栈溢出等。这些错误比较严重,仅靠程序本身很难解决,因此程序一般不对其进行处理。

2. Exception 类

Exception 类为异常类,是程序可以处理的异常。在 Java 语言中程序进行的异常处理,都属于 Exception 类。Exception 类的子类分为两种类型:运行时异常和非运行时异常。

运行时异常是 RuntimeException 类及其子类异常,如 NullPointerException、IndexOutOfBoundsException 等。这些异常也称为免检异常。在程序编写过程中,对这类异常,即使不编写异常处理代码,依然可以通过编译,只是在程序运行时可能报错,因此称之为运行时异常或者免检异常。运行时异常一般由程序逻辑错误引起,因此避免这类异常的发生应该从逻辑角度去考虑。

非运行时异常是指 RuntimeException 以外的异常,如 IOException、ClassNotFoundException 等。在程序编写过程中,必须对该类异常进行处理,否则程序就不能编译通过。

常见的异常类型如表 8.1 所示。

表 8.1 常见异常类型及其说明

异 常 类 型	说 明
Exception	异常层次结构的根类
RuntimeException	运行时异常
ArithmeticException	算术异常,如零作为除数
ArrayIndexOutOfBoundException	数组角标越界异常
NullPointerException	空引用异常
NumberFormatException	数字转化格式异常
IOException	I/O 异常的根类

续表

异 常 类 型	说 明
FileNotFoundException	找不到文件异常
EOFException	文件结束
InterruptedException	线程中断异常
ClassCastException	对象转换异常
SQLException	数据库操作异常

8.1.2 异常的捕获和处理

Java 语言中,异常处理机制为捕获异常和抛出异常。异常捕获通常使用 try-catch 语句,一般语法格式如下:

```
try {
    // 可能会发生异常的程序代码
} catch (ExceptionType id){
    // 捕获并处置异常
}
```

上述代码中,try{}中的语句块包含可能产生异常的代码段,catch(){}中的语句块用来对捕获到的异常进行处理。一个 try 语句块中可以捕获多个异常,并对不同类型的异常做出不同的处理。如果单个 try 语句块引发多个异常,catch 语句可以有多个,用来处理不同类型的异常。当异常被引发时,依次与 catch 语句块中声明的异常类型进行匹配,当某一个匹配异常类型的语句被执行时,后面的 catch 语句块将不会被执行。因此,有多个 catch 语句块时,异常类型的排列顺序应符合子类异常在前,父类异常在后的原则。

【例 8.2】 常见异常处理示例。

针对例 8.1 的异常进行处理,代码如下:

```
public class Main {
    public static void main(String[ ] args) {
        try {
            HuskyDog dog = null;
            dog.moveUp();
        }catch(NullPointerException e) {
            e.printStackTrace();
        }
        Screen screen = new Screen(8);
        HuskyDog dog1 = new HuskyDog(4,5);
        screen.add(dog1);
    }
}
```

视频讲解

运行程序,屏幕正常显示出来。在 try{}代码块中发生了异常,程序转而执行 catch(){}中的代码,将异常原因打印出来。对异常处理完毕后,程序仍会向下执行,不会因为异常终止运行。try{}代码块中,在发生异常的语句之后的代码不会被执行。如果程序中需要确保有些语句必须要被执行,如释放资源、关闭线程池等,可以在 try-catch 语句后,加上一个 finally{}代码块。finally{}是可选项,无论异常产生与否,finally{}中的语句块通常情况下都会被执行,即使在try-catch语句中使用了"return"语句也不例外。唯一例外的是,在try-catch语句中执行了"System.exit(0)"语句终止程序运行,finally{}中的语句块将不会被执行。

8.1.3 异常的抛出

有时,方法中出现异常,但当前作用域没有能力处理这个异常,则将该异常向上抛出,交由上层的作用域来处理。如同生活中遇到有人突然晕倒,情况特别严重的时候,就需要拨打"120"急救电话,请专业人员来处理。

Java语言抛出异常通常使用throws或throw关键字。

1. 使用throws抛出异常

如果一个方法可能会出现异常,但没有能力处理这种异常,可以在方法声明处用throws来声明抛出异常,然后让调用者在使用时再进行异常处理。使用throws抛出异常的基本格式如下:

```
[修饰符] 返回值类型 方法名(参数列表) throws 异常类列表{
    //方法体
}
```

视频讲解

throws关键字后面跟的是异常类,可以一个或者多个,多个时用逗号隔开。

【例8.3】 使用throws关键字抛出异常示例。

```java
public class Main {
    public static void play(HuskyDog dog) throws NullPointerException{
        dog.moveUp();
    }
    
    public static void main(String[] args) {
        HuskyDog dog = null;
        try {
            play(dog);
        }catch(NullPointerException e) {
            e.printStackTrace();
        }
        
        Screen screen = new Screen(8);
    }
}
```

运行程序,"屏幕"正常显示,并且控制台输出异常提示信息,如图8.3所示。

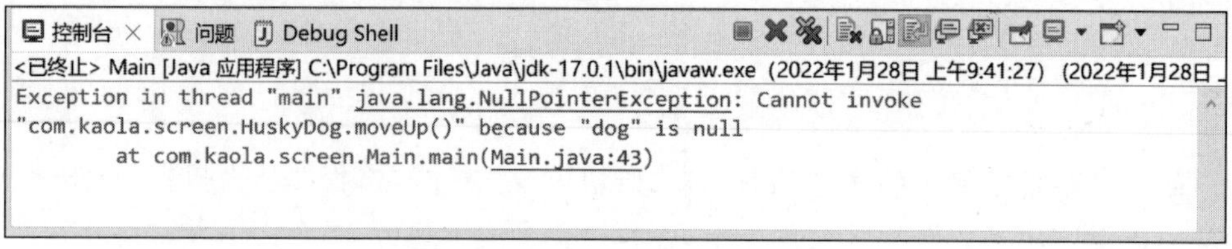

图8.3 异常提示信息

需要注意的是，调用有抛出异常的方法时，除了可以在调用程序中直接进行处理，也可以使用 throws 关键字继续将异常抛出。这与生活中处理复杂问题情况相似，如果当前级别解决不了的问题，就向上汇报，如果上级也处理不了，就继续上报，直到能够处理为止。

2. 使用 throw 抛出异常

throw 关键字也能抛出异常，与 throws 不同的是，throw 在方法体内使用，抛出的不是异常类，而是一个异常实例，而且每次只能抛出一个异常实例。使用 throw 关键字抛出异常的基本格式如下：

```
throw throwableInstance;
```

throwableInstance 是 Throwable 类或者其子类的实例，可以是用户创建或者程序捕捉到的异常对象。

【例 8.4】 使用 throw 关键字抛出异常示例。

在 play() 方法中创建一个异常对象并将其抛出，代码如下：

```java
public class Main {
    public static void play(HuskyDog dog) {
        if(dog == null) {
            throw new NullPointerException("对象为空");
        }
        else {
            dog.moveUp();
        }
    }

    public static void main(String[ ] args) {
        HuskyDog dog = null;
        try {
            play(dog);
        }catch(NullPointerException e) {
            e.printStackTrace();
        }
        
        Screen screen = new Screen(8);
    }
}
```

运行程序，"屏幕"正常显示，并且控制台输出相应异常提示信息，如图 8.4 所示。

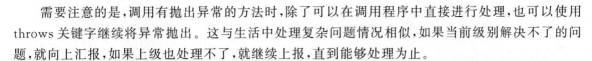

图 8.4 异常提示信息

采用 throw 关键字抛出异常之后，本次方法的调用即告结束，与 return 语句相似。

8.1.4 自定义异常

尽管 Java 语言中定义了大量的异常类，但对于某些特殊的异常情况，还需要创建自定义的异常类进行处理。自定义的异常类必须继承自 Exception 类或其子类。

视频讲解

【例 8.5】 创建自定义异常类，用于提示游戏角色越界。

自定义屏幕越界类，代码如下：

```java
public class ScreenOutOfBoundException extends Exception{
    public ScreenOutOfBoundException() {
        super();
    }

    public ScreenOutOfBoundException(String message) {
        super(message);
    }
}
```

设计"泰迪"类，如果设置的位置超出了屏幕范围，就抛出相应的异常，代码如下：

```java
public class TeddyDog extends Dog{
    public TeddyDog(int r, int c) throws ScreenOutOfBoundException {
        super(r, c);
        if(r < 0 || r > 7 || c < 0 || c > 7) {
            throw new ScreenOutOfBoundException("超出屏幕范围");
        }
    }
}
```

测试代码如下：

```java
public class Main {
    public static void main(String[] args) {
        try {
            TeddyDog dog = new TeddyDog(8,0);
        } catch (ScreenOutOfBoundException e) {
            e.printStackTrace();
        }
        Screen screen = new Screen(8);
    }
}
```

运行程序，"屏幕"正常显示，控制台输出异常提示信息，如图 8.5 所示。

```
控制台 ×  问题  Debug Shell
<已终止> Main [Java 应用程序] C:\Program Files\Java\jdk-17.0.1\bin\javaw.exe (2022年1月28日 下午9:59:27) (2022年1月28日
com.kaola.screen.ScreenOutOfBoundException: 超出屏幕范围
        at com.kaola.screen.TeddyDog.<init>(TeddyDog.java:8)
        at com.kaola.screen.Main.main(Main.java:9)
```

图 8.5 异常提示信息

8.2 综合案例：重构"飞机大战"游戏

视频讲解

异常并不可怕，建立应对异常的机制非常重要。设计程序，完善"飞机大战"游戏，保证按键控制的飞机不能飞出屏幕的边界。

利用例 8.5 设计的自定义异常类修改飞机类中控制飞机运动的方法，如果操作会导致飞机越界，则抛出相应的异常，代码如下：

```java
public class Plane extends GameObject implements Move{
    public Plane(int r, int c) {
        super(r,c);
    }

    @Override
    public void moveUp() {
        try {
            if(getRow() - 1 < 0) {
                throw new ScreenOutOfBoundException("超出屏幕范围");
            }
            else {
                setRow(getRow() - 1);
            }
        }catch(ScreenOutOfBoundException e){
            e.printStackTrace();
        }
    }

    @Override
    public void moveDown() {
        try {
            if(getRow() >= Config.SCREENSIZE - 1) {
                throw new ScreenOutOfBoundException("超出屏幕范围");
            }
            else {
                setRow(getRow() + 1);
            }
        }catch(ScreenOutOfBoundException e){
            e.printStackTrace();
        }
    }

    @Override
    public void moveLeft() {
        try {
            if(getCol() - 1 < 0) {
                throw new ScreenOutOfBoundException("超出屏幕范围");
            }
            else {
                setCol(getCol() - 1);
```

```java
            }
        }catch(ScreenOutOfBoundException e){
            e.printStackTrace();
        }
    }

    @Override
    public void moveRight() {
        try {
            if(getCol() >= Config.SCREENSIZE - 1) {
                throw new ScreenOutOfBoundException("超出屏幕范围");
            }
            else {
                setCol(getCol() + 1);
            }
        }catch(ScreenOutOfBoundException e){
            e.printStackTrace();
        }
    }

    public Bullet shoot() {
        Bullet bullet = new Bullet(getRow(),getCol());
        return bullet;
    }

    public class Bullet extends GameObject {
        public Bullet(int row, int col) {
            super(row,col);
        }
        public void moveUp() {
            setRow(getRow() - 1);
        }
    }
}
```

测试类不需要修改，代码如下：

```java
public class Main {
    public static void main(String[] args){
        ArrayList<Plane.Bullet> bullets = new ArrayList<Plane.Bullet>();
        Screen screen = new Screen(Config.SCREENSIZE);
        Plane plane = new Plane(7,4);
        Plane.Bullet bullet = null;
        screen.add(plane);

        char key;
        while(true) {
            screen.delay();
            key = screen.getKey();

            if(key == 'w') {
                plane.moveUp();
            }
```

```
            if(key == 's') {
                plane.moveDown();
            }
            if(key == 'a') {
                plane.moveLeft();
            }
            if(key == 'd') {
                plane.moveRight();
            }

            if(key == 'k') {
                bullet = plane.shoot();
                screen.add(bullet);
                bullets.add(bullet);
            }

            for(Plane.Bullet blet:bullets) {
                blet.moveUp();
            }
        }
    }
}
```

运行程序,按键控制"飞机"上、下、左、右运动,"飞机"不能飞出屏幕边界。

习题

8.1 关于异常的含义,下列描述中正确的是_____。
   A. 程序编译错误   B. 程序语法错误
   C. 程序自定义的异常事件   D. 程序编译或运行时发生的异常事件

8.2 以下对异常的描述不正确的有_____。
   A. 异常分为 Error 和 Exception
   B. Throwable 是所有异常类的父类
   C. Exception 是所有异常类父类
   D. Exception 包括 RuntimeException 和 RuntimeException 之外的异常

8.3 对于已经被定义过可能抛出异常的语句,在编程时_____。
   A. 必须使用 try-catch 语句处理异常,或用 throws 将其抛出
   B. 如果程序错误,必须使用 try-catch 语句处理异常
   C. 可以置之不理
   D. 只能使用 try-catch 语句处理

8.4 以下对自定义异常描述正确的是_____。
   A. 自定义异常必须继承自 Exception
   B. 自定义异常可以继承自 Error
   C. 自定义异常可以更加明确定位异常出错的位置和给出详细出错信息
   D. 程序中已经提供了丰富的异常类,使用自定义异常没有意义

第9章 字符串

字符串类是Java语言中常用的类,用于处理各种字符串。字符串是一连串的字符,可以包含字母、数字和其他符号。Java语言提供了三个字符串类:String类、StringBuffer类和StringBuilder类。其中,String类用于处理不变的字符串,StringBuffer类和StringBuilder类用于处理可变的字符串。

9.1 String类

String类用于处理不变字符串,是最常用的字符串类。

9.1.1 创建String类对象

创建String对象的方法有如下两种。
(1) 使用字符串常量创建字符串对象。例如:

```
String s = "Hello,world";
```

(2) 使用String类的构造方法创建字符串对象。例如:

```
String s = new String("Hello,world");
```

String类有11个重载的构造方法,可以由指定的字符串、字符数组或者字节数组生成字符串,常用的构造方法如表9.1所示。

表 9.1 String 类的构造方法及其功能

方　法	功　能
String()	创建一个空字符串
String(char[] value)	使用字符数组创建字符串对象
String(byte[] value)	使用字节数组创建字符串对象
String(String value)	使用字符串创建字符串对象

9.1.2　字符串类常用方法

String 类的常用方法如表 9.2 所示。

表 9.2 String 类的常用方法及其功能

方　法	功　能
int length()	获取一个字符串的字符个数
char charAt(int index)	从字符串中取出指定位置的字符
int indexOf("字符")	查找一个指定的字符串是否存在,返回的是字符串的位置,如果不存在,则返回－1
boolean equals(Object anObject)	比较两个字符串是否相等
boolean startsWith(String prefix)	判断此字符串是否以指定的字符串开始
boolean endsWith(String suffix)	判断此字符串是否以指定的字符串结束
boolean contains(CharSequence chars)	判断是否包含指定的字符系列
String substring(int beginIndex)	返回一个字符串,该字符串是此字符串的子字符串。子字符串从指定的 beginIndex 开始,直到此字符串末尾的所有字符
String[] split(String regex)	根据给定正则表达式的匹配拆分此字符串

【例 9.1】　编写程序,实现密码长度的检测。登录系统中,密码的长度一般要求为 6 个字符,如果输入密码的长度不符合要求,给出相应的提示。

判断字符串的长度,需要使用到 String 类的 length()方法,代码如下:

```java
public class Main{
    public static void main(String[] args) {
        String password = JOptionPane.showInputDialog("请输入密码:");
        if(password.length() != 6) {
            JOptionPane.showMessageDialog(null,"输入密码的长度不对");
        }
    }
}
```

视频讲解

运行程序,在对话框中输入密码,当输入密码长度不为 6 的时候,会弹出相应的提示。JOptionPane 类是 Java 语言提供的消息提示对话框类,其中 showInputDialog()方法是显示输入对话框,showMessageDialog()方法是显示消息对话框。

【例 9.2】　编写程序,实现登录功能。输入用户名和密码,当用户名和密码都正确的时候,提示登录成功,否则给出相应的提示,用户名为 Lilei,密码为 123456。

通过 JOptionPane 类中的 showInputDialog()方法弹出输入对话框,输入用户名和密码,然后根据输入信息进行判断登录是否成功。登录成功的条件是输入的用户名字符串和密码字符串要与设置好的字符串相等。判断字符串相等,需要使用 String 类的 equals()方法,代码如下:

视频讲解

```java
public class Main {
    public static void main(String[ ] args) {
        String userName = JOptionPane.showInputDialog("请输入用户名:");
        String password = JOptionPane.showInputDialog("请输入密码:");

        if(userName.equals("Lilei") && password.equals("123456")) {
            JOptionPane.showMessageDialog(null,"信息正确,登录成功");
        }

        if(userName.equals("Lilei") == false ) {
            JOptionPane.showMessageDialog(null,"用户名不正确,登录失败");
        }

        if(password.equals("123456") == false ) {
            JOptionPane.showMessageDialog(null,"密码不正确,登录失败");
        }
    }
}
```

运行程序,在对话框中输入用户名和密码,会弹出相应的提示。当用户名和密码都正确的时候,会提示登录成功;当用户名或者密码输入不正确的时候,会给出相应的登录失败的提示。

使用 String 类,需要注意的是 String 类是 final 类型,一旦创建了 String 对象,其内容就不能改变。例如:

```java
String s = "Hello";
String s1 = s.toUpperCase();
```

toUpperCase()方法是字符串中所有的小写字母都转换成大写。字符串 s1 的值为"HELLO",而 s 的值仍然是"Hello"。toUpperCase()方法是产生了一个新的字符串,而原字符串的内容不会改变。在实际应用中,有时候需要对字符串进行修改,Java 语言提供了 StringBuffer 和 StringBuilder 类来操作字符串,StringBuffer 类和 StringBuilder 类定义的是可变字符串,其对象内容可以被修改。

9.2 StringBuffer 类和 StringBuilder 类

StringBuffer 类和 StringBuilder 类功能相似,两个类中所提供的方法基本相同,它们的主要区别在于 StringBuilder 类的方法没有实现线程安全,处理字符串时有速度优势。StringBuffer 类的常用方法如表 9.3 所示。

表 9.3 StringBuffer 类的常用方法及其功能

方　　法	功　　能
StringBuffer append(String s)	在当前字符串末尾添加字符串
StringBuffer insert(int offset,String str)	在指定位置插入字符串
StringBuffer delete(int start,int end)	删除从 start 开始到 end(不包括 end)之间的字符

续表

方法	功能
StringBuffer reverse()	将此字符序列用其反转形式取代
StringBuffer replace(int start,int end,String str)	使用给定 String 中的字符替换此序列的子字符串中的字符

视频讲解

【例 9.3】 编写程序，实现信息保护功能。为了保护用户的信息安全，通常显示用户的手机号码，会将中间的 4 位用"＊"替代。

字符串的内容发生了改变，所以需要 StringBuffer 类或者 StringBuilder 类生成字符串对象。字符串中间的字符被替换，则需要使用 replace()方法，代码如下：

```java
public class Main {
    public static void main(String[ ] args) {
        String num = JOptionPane.showInputDialog("请输入手机号码:");
        StringBuilder showNum = new StringBuilder(num);
        showNum.replace(3, 7, "****");
        showNum.insert(0, "手机号码为:");
        JOptionPane.showMessageDialog(null,showNum);
    }
}
```

运行程序，在对话框中输入手机号码，显示的时候会将中间 4 位用"＊"代替。

9.3 综合案例：数据加密和解密

视频讲解

信息安全的重要性不言而喻，大到国家军事、政治机密，小到个人各种账户信息，都需要严格保护。在网络中传输的信息通常要经过加密处理之后传送。设计系统的时候，为了保护用户的隐私，存储用户密码通常采用 MD5 加密方式存储，即使用户信息数据库被窃取，依然能够保护用户的隐私。MD5 算法用于密码保护的原理是用户登录时，系统根据用户输入的密码计算出 MD5 值，然后去和系统中保存的 MD5 值进行比较，如果一致则意味着密码正确。除非知道原始密码，否则即使查看源代码或者数据库存储的密码信息，也无法逆向得到原始密码，这样就能保证用户隐私。

Java 语言提供了 MessageDigest 加密工具类用于字符串加密，使用时先将字符串转换成字节数组，再进行 MD5 算法运算，结果仍为字节数组，然后将字节数组转换成字符串，代码如下：

```java
public class Main {
    public static void main(String[ ] args) {
        String password = JOptionPane.showInputDialog("请输入密码:");
        String SysPassWord = "4QrcOUm6Wau + VuBX8g + IPg == ";      //加密后的密码

        if(SysPassWord.equals(md5(password))) {
            JOptionPane.showMessageDialog(null,"密码正确,登录成功");
        }
        else {
            JOptionPane.showMessageDialog(null,"密码不正确,登录失败");
        }
    }
```

```
private static String md5(String password) {                    //MD5 加密算法
    try {
        MessageDigest md = MessageDigest.getInstance("md5");
        byte[] bytes = md.digest(password.getBytes());          //获得密文完成的哈希运算
        String str = Base64.getEncoder().encodeToString(bytes); //转换成字符串
        return str;
    } catch (NoSuchAlgorithmException e) {
        e.printStackTrace();
    }
    return null;
}
```

运行程序,在对话框中输入密码"123456",登录成功,否则失败。如果不知道密码是123456,即使查看源代码,也不知道原始密码,这样就能保护用户隐私。

习题

9.1 执行下列程序段之后,字符串 str 的值为_____。

```
String str = "Hello";
str.replace('o', 'a');
str.concat("World");
```

  A. Hello   B. HellaWorld C. HelloWorld D. Hella

9.2 执行下列程序段之后,字符串 str 的值为_____。

```
StringBuffer str = new StringBuffer("Hello");
str.replace(4,5,"a");
str.append("World");
```

  A. Hello   B. HellaWorld C. HelloWorld D. Hella

9.3 编写加密程序,实现输入原始字符串,输出加密字符串。加密规则是按照英文字母表的顺序,将每个字母转换为下一个字母表示,如果原始字符为"a",则转换为"b";原始字符为"A",则转换为"B";如果原始字符为"z",则转换为"a"。

9.4 编写解密程序,实现输入加密字符串,输出原始字符串。加密规则是按照英文字母表的顺序,将对每个字母转换为下一个字母表示,如果原始字符为"a",则转换为"b";原始字符为"A",则转换为"B";如果原始字符为"z",则转换为"a"。

第10章

输入输出

输入输出(I/O)是指计算机与外部设备进行数据传递的操作。几乎所有的程序都具有输入与输出操作。例如，从键盘上读取数据、将数据输出到屏幕上、从本地或网络上的文件读写数据等。程序的输入和输出可以说是程序与用户之间沟通的桥梁，通过输入和输出操作可以从外界接收信息，或者是把信息传递给外界，实现用户和程序之间的交互。Java 语言把这些输入与输出操作用流来实现，通过统一的接口来表示，使输入输出相关的程序设计变得简单。

10.1 流的概念

Java 语言的输入输出采用流来实现，流(Stream)是指在计算机的输入输出操作中各部件之间的数据流动的序列，由字节组成。数据在传递的过程中，就像源源不断的水流一样，故而形象地被称为"流"。

数据流是 Java 程序发送和接收数据的一个通道，按照数据的传输方向，数据流可分为输入流与输出流。输入流就是把来自外部设备的数据读入程序中，输出流就是把程序中的数据输出外部设备或其他地方。从文件角度简单总结就是，输入流用来读取数据，输出流用来写入数据。在 Java 程序中，要想从文件中读取数据，需要在程序和文件之间建立一条数据输入的通道。反之，要将数据写入文件中，也需要在程序和文件之间建立一条数据输出的通道。当程序创建输入或输出流对象时，Java 会自动建立数据输入或输出的通道，如图 10.1 所示。

输入输出流都是单向的，输入流只能用来读取数据，而不能用来写入数据；输出流只能用来写入数据，而不能用来读取数据。在一个数据传输通道中，如果既要写入数据，又要读取数据，则要分别提供输入流和输出流。

流序列中的数据既可以是未经加工的原始二进制数据，也可以是经过一定编码处理后符合某种格式规定的特定数据。因此，流又分为字节流和字符流两种。

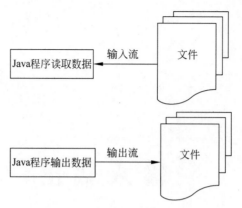

图 10.1 程序与文件间的数据通道

(1) 字节流。数据流中最小的数据单元是字节,用于读取或写入二进制数据。

(2) 字符流。数据流中最小的数据单元是字符,用于读取或写入字符数据。Java 语言中的字符是 Unicode 编码,1 个字符占用 2 字节。

在计算机系统底层,所有的输入输出都是字节形式,字符流只为处理字符提供方便有效的方法。在 Java 语言中使用字节流和字符流的步骤基本相同,首先创建一个与数据相关的流对象,然后利用流对象的方法输入或者输出数据,最后使用 close()方法关闭流。

10.2 字节流

10.2.1 InputStream 类和 OutputStream 类

字节流的最顶层是两个抽象类:InputStream 类和 OutputStream 类,其他处理字节的输入输出类都是它们的子类。这些子类应用于不同的外设,如磁盘文件、网络连接、内存缓冲区等。抽象类 InputStream 和 OutputStream 中定义了关键方法,分别是 read()方法和 write()方法,用于对数据的字节进行读和写。这两种方法都是抽象方法,需要被许多子类重载,这些子类如图 10.2 所示。

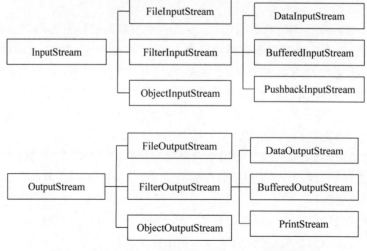

图 10.2 InputStream 类和 OutputStream 类及其子类

InputStream类和OutputStream类中的方法定义都抛出了IOException异常,使用这些方法时需要使用try-catch结构捕获异常,或者声明方法抛出异常。

10.2.2 字节流读写文件

计算机的数据基本保存在硬盘文件中,对文件的读写是常见操作。针对文件的读写操作,Java提供了两个专门的类,分别是FileInputStream类和FileOutputStream类。FileInputStream类是InputStream类的子类,操作文件的字节输入流,专门用于读取文件中的数据。FileOutputStream类是OutputStream类的子类,操作文件的字节输出流,专门用于将数据写入文件。

【例10.1】 编写程序,复制Chapter 10项目中images文件夹中的"Box.png"图片文件,复制的文件名为"BoxCopy.png"。

视频讲解

复制文件需要完成两个步骤:获得数据和写入数据。因此,需要建立输入流和输出流对象,先从输入流中读取数据,然后向输出流中写入数据。在复制的过程中,使用while循环语句,逐字节地进行读取和写入,直到将文件所有的字节都复制到指定文件中,代码如下:

```java
public class Main {
    public static void main(String[ ] args) throws Exception{
        FileInputStream in = new FileInputStream("images/Box.png");
        FileOutputStream out = new FileOutputStream("images/BoxCopy.png");
        int len = 0;
        while((len = in.read()) != -1) {
            out.write(len);
        }
        in.close();
        out.close();
    }
}
```

运行程序,在images文件夹下出现"BoxCopy.png"文件,实现了文件的复制。

复制的主要过程是通过文件名来创建FileInputStream类和FileOutputStream类的对象,通过输入流的read()方法每次从输入流中读取1字节,然后通过输出流的write()方法将该字节写入指定文件,直到达到文件末尾就返回"-1",结束循环,完成文件的复制。最后,调用close()方法关闭这些字节流。

10.2.3 缓冲字节流读写文件

逐字节地读写文件,需要频繁地操作文件,效率非常低。为了提高读写效率,可以先定义一个数组作为缓冲区,这样一次就能读写多字节的数据,代码如下:

```java
public class Main {
    public static void main(String[ ] args) throws Exception{
        FileInputStream in = new FileInputStream("images/Box.png");
        FileOutputStream out = new FileOutputStream("images/BoxCopy1.png");
        int len = 0;
        byte [ ] buff = new byte[1024];
        while((len = in.read(buff)) != -1) {
```

```
            out.write(buff,0,len);
        }
        in.close();
        out.close();
    }
}
```

使用缓冲区读写文件,可以减少对文件的操作次数,提高读写数据的效率。除了自己定义缓冲区外,系统还提供了两个带缓冲的字节流,分别是 BufferedInputStream 类和 BufferedOutputStream 类,在读写数据的时候提供缓冲功能。BufferedOutputStream 类的操作原理是:向缓冲流写入数据时,数据先写到缓冲区,待缓冲区写满后,将缓冲区的数据一次性发送给输出设备。BufferedInputStream 类的操作原理是:从缓冲流中读取数据时,系统先从缓冲区读出数据,待缓冲区为空时,系统再从输入设备读取数据到缓冲区。

这两个类用来将其他的字节流包装成缓冲字符流,以提高读写数据的效率。使用缓冲字节流复制文件的代码如下:

```
public class Main {
    public static void main(String[ ] args) throws Exception{
        FileInputStream in = new FileInputStream("images/Box.png");
        FileOutputStream out = new FileOutputStream("images/BoxCopy2.png");
        BufferedInputStream bis = new BufferedInputStream(in);
        BufferedOutputStream bos = new BufferedInputStream(out);
        int len = 0;
        while((len = bis.read()) != -1) {
            bos.write(len);
        }
        in.close();
        out.close();
        bis.close();
        bos.close();
    }
}
```

运行程序,在 images 文件下出现"BoxCopy2.png"文件,实现了文件的复制。BufferedInputStream 类和 BufferedOutputStream 类创建流对象时,会创建一个默认大小的内置缓冲区数组,通过缓冲区进行读写,减少系统输入输出读取次数,从而提高读写的效率。

10.3 字符流

程序中涉及字符操作时,使用字符流会比字节流更加合适。字符流的最顶层是两个抽象类:Reader 类和 Writer 类。Reader 类是字符输入流,用于从设备中读取字符。Writer 类是字符输出流,用于向设备写入字符。其他关于处理字符的类都是它们的子类,这些子类对不同的外设进行处理,如磁盘文件、网络连接、内存缓冲区等。字符流类的继承关系如图 10.3 所示。

字符流类的继承关系与字节流类的继承关系较为相似,许多子类都是成对出现。

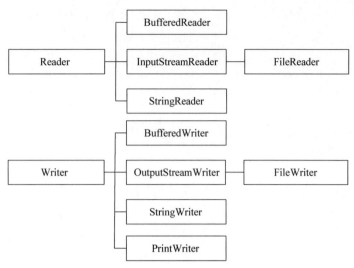

图 10.3 Reader 类和 Writer 类及其子类

10.3.1 字符流读写文件

字符流中读写文件的类是 FileReader 类和 FileWriter 类,FileReader 类是以字符的形式读取文件的内容,FileWriter 类是向文件写入字符,使用方式与 FileInputStream 类和 FileOutputStream 类相似。

【例 10.2】 编写程序,复制 Chapter 10 项目中 configuration 文件夹下的"Configuration.txt"文本文件,复制的文件名为"ConfigurationCopy.txt"。

复制文本文件,使用字符流更为合适,代码如下:

视频讲解

```
public class Main {
    public static void main(String[ ] args) throws Exception{
        FileReader in = new FileReader("configuration/configuration.txt");
        FileWriter out = new FileWriter("configuration/configurationCopy.txt");
        int len = 0;
        char [ ] buff = new char[1024];
        while((len = in.read(buff)) != -1) {
            out.write(buff,0,len );
        }
        in.close();
        out.close();
    }
}
```

运行程序,在 configuration 文件夹下出现"ConfigurationCopy.txt"文件,并且内容与"Configuration.txt"文件一致。

10.3.2 字符缓冲流读写文件

字符流与字节流类似,也提供了字符缓冲流读写文件,分别是 BufferedReader 类和 BufferedWriter 类。这两个类实现了具有缓冲功能的字符输入输出流,用来将其他的字符流包装

成缓冲字符流,以提高读写数据的效率。使用缓冲字符流复制文件的代码如下:

```java
public class Main {
    public static void main(String[] args) throws Exception{
        FileReader in = new FileReader("configuration/configuration.txt");
        FileWriter out = new FileWriter("configuration/configurationCopy1.txt");

        BufferedReader br = new BufferedReader(in);
        BufferedWriter bw = new BufferedWriter(out);
        int len = 0;
        while((len = br.read()) != -1) {
            bw.write(len);
        }
        in.close();
        out.close();
        br.close();
        bw.close();
    }
}
```

BufferedRead 类中还有一个 readLine()方法,该方法用于一次读取一行文本。使用该方法实现文件的复制,代码如下:

```java
public class Main {
    public static void main(String[] args) throws Exception{
        FileReader in = new FileReader("configuration/configuration.txt");
        FileWriter out = new FileWriter("configuration/configurationCopy2.txt");

        BufferedReader br = new BufferedReader(in);
        BufferedWriter bw = new BufferedWriter(out);
        int len = 0;
        String str = null;
        while((str = br.readLine()) != null) {
            bw.write(str);
            bw.newLine();                  // 将下一行分隔成新的一行
        }
        in.close();
        out.close();
        br.close();
        bw.close();
    }
}
```

运行程序,文件复制成功。

10.4 标准输入输出流

计算机系统都有标准的输入输出设备,通常情况下,键盘是标准输入设备,屏幕是标准输出设备。Java 程序经常需要使用这些设备,因此,Java 系统事先定义了两个对象 System.in 和 System.

out,分别与系统的标准输入输出相联系。

System.in 是标准输入流,是 InputStream 类的对象,可以使用 read()方法从键盘上读取字节,也可以将其包装成字符流用来读取各种类型的数据。System.out 是标准输出流,是 PrintStream 类的对象,能够输出各种类型的数据,输出设备默认是控制台,在程序中经常会使用到。

【例 10.3】 编写程序,读取 Chapter 10 项目中 configuration 文件夹目录下"Configuration.txt"文本文件,并将内容显示在控制台上。

将内容显示在控制台上,可以使用标准输出流,代码如下:

```java
public class Main {
    public static void main(String[ ] args) throws Exception{
        FileReader in = new FileReader("configuration/configuration.txt");
        BufferedReader br = new BufferedReader(in);
        String str;
        while((str = br.readLine()) != null) {
            System.out.println(str);
        }
        in.close();
    }
}
```

视频讲解

程序运行结果如图 10.4 所示。

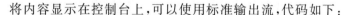

图 10.4 标准输出流运行结果

【例 10.4】 编写程序,实现猜数字游戏。

猜数字游戏是一种古老的密码破译类益智小游戏,程序随机产生 0~99 中任意一个整数,用户通过键盘输入数字进行猜测。如果猜对了,程序提示猜对了;如果猜错了,程序会提示猜大了或者猜小了。最多给予 10 次机会,如果都猜错了,程序提示游戏结束,并输出正确数字。代码如下:

```java
public class Test{
    public static void main(String[ ] args) throws IOException{
        System.out.println("请输入 0~100 中的任意一个整数");
        BufferedReader br = new BufferedReader(new InputStreamReader(System.in));
        Random rand = new Random();
        int num = rand.nextInt(100);
        int key = 0;
        String str = null;
        int time = 10;
        while(time > 0){
            str = br.readLine();
            key = Integer.parseInt(str);
```

视频讲解

```
        if(key == num){
            System.out.println("恭喜您,答对了");
            break;
        }else if(key > num){
            System.out.println("您猜的数字大了");
        }else{
            System.out.println("您猜的数字小了");
        }
        time--;
        System.out.println("您还有" + time + "次机会");
    }

    if(time == 0){
        System.out.println("游戏结束,未能猜对数字,正确的数字为" + num);
    }
    br.close();
    }
}
```

运行程序,在控制台输入数据,猜测随机产生的数字。在程序中,用到了 Random 随机类的 nextInt()方法产生某一范围内的随机数,并且用到了 Integer 类的静态成员方法 parseInt()方法将字符串转换成整数。对于这些常用类的使用,读者可以查阅相应的资料学习。

10.5 对象序列化

在程序中,有时需要将数据长期保存下来。例如,观看视频的时候,会将数据信息保存下来,以便下次继续观看。要将对象中的数据长期保存,就需要使用对象序列化。对象序列化是指将内存中保存的对象以数据流的形式进行处理和传输,可以实现对象的保存或网络传输。

Java 语言提供了一种对象序列化的机制,使内存中的 Java 对象转换成与平台无关的二进制数据流,这种二进制数据流既可以长期保存在磁盘上,又可以通过网络传输到另一个网络节点。将程序中的对象转换为字节序列的过程,称为对象序列化。反之,从字节序列恢复为对象的过程称为反序列化。

一个类的对象序列化成功的前提是该类必须实现 Serializable 接口或者 Externalizable 接口。由于 Serializable 接口简单易用,实际开发中大多数程序都采用 Serializable 接口实现序列化。Serializable 接口中没有定义任何方法,不需要实现任何方法,只是为了标注该对象是可被序列化的。实现对象序列化的具体过程,需要依靠 ObjectOutputStream 类和 ObjectInputStream 类。ObjectOutputStream 类可以将对象转换为特定格式的二进制数据输出,ObjectInputStream 类可以读取二进制数据,并将其转换为具体类型的对象。

对象序列化的过程是:创建一个 ObjectOutputStream 输出流对象,然后调用该对象的 writeObject()方法输出序列化对象。反序列化的过程是:创建一个 ObjectInputStream 输入流对象,然后调用该对象的 readObject()方法输出对象。

视频讲解

【例 10.5】 编写程序,将"飞机大战"游戏中"飞机"对象保存到项目目录 data 文件夹中。
如果想将"飞机"对象序列化,则需要"飞机"类实现 Serializable 接口,飞机 Plane 类继承

GameObject 类，所以直接将 GameObject 实现 Serializable 接口，代码如下：

```java
public class GameObject implements Serializable{
    private int row;
    private int col;
    private int size = 1;
    public GameObject() {
    }

    public GameObject(int r, int c) {
        row = r;
        col = c;
    }

    public int getRow() {
        return row;
    }
    public void setRow(int row) {
        this.row = row;
    }

    public int getCol() {
        return col;
    }
    public void setCol(int col) {
        this.col = col;
    }

    public int getSize() {
        return size;
    }
    public void setSize(int size) {
        this.size = size;
    }
}
```

飞机类则不需要修改。

测试代码如下：

```java
public class Main {
    public static void main(String[ ] args) throws Exception{
        Screen screen = new Screen(8);
        Plane plane = new Plane(5,7);
        screen.add(plane);
        FileOutputStream os = new FileOutputStream("data/objectFile.obj");
        ObjectOutputStream out = new ObjectOutputStream(os);

        out.writeObject(plane);
        out.close();
        os.close();
    }
}
```

运行程序，对应文件夹下出现"objectFile.obj"文件，至于保存的数据是否正确，需要通过例 10.6 中的代码进行检测。

【例 10.6】 编写程序,读取保存的"飞机大战"游戏中"飞机"对象数据,并根据数据恢复游戏。本例题需要将例 10.5 保存的数据,反序列化成具体对象,测试代码如下:

```
public class Main {
    public static void main(String[ ] args) throws Exception{
        Screen screen = new Screen(8);
        FileInputStream is = new FileInputStream("data/objectFile.obj");
        ObjectInputStream in = new ObjectInputStream(is);

        Plane plane = (Plane) in.readObject();          //读取飞机对象
        screen.add(plane);
        in.close();
        is.close();
    }
}
```

运行程序,根据保存的数据直接在屏幕对应位置显示出"飞机"。经过对比可知,飞机的位置与保存前的一致。

10.6 综合案例:游戏数据的存档和读取

编写程序,实现按键"P"保存游戏数据,对游戏进行存档。下次重新玩游戏的时候,直接读取数据,恢复游戏,如图 10.5 所示。

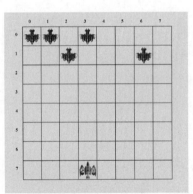

图 10.5 "飞机大战"

在"飞机大战"游戏类 PlaneGame 中新增两个方法,一个是保存数据的 saveData()方法,另外一个是读取数据的 readData()方法。保存和读取的数据就是屏幕上所有元素的数据信息,获得屏幕上所有元素的信息可以通过 Screen 类的 getData()方法,代码如下:

```
public class PlaneGame extends GameCore{
    private Screen screen;
    Plane plane;
    HashSet< Plane.Bullet > bullets = new HashSet< Plane.Bullet >();
    Plane.Bullet bullet;
```

```java
public void saveData(ArrayList<GameObject> objectList) {
    ObjectOutputStream out;
    try {
        out = new ObjectOutputStream(new FileOutputStream("data/ objectFileAll.obj"));
        out.writeObject(objectList);
        out.close();
    } catch (FileNotFoundException e) {
        e.printStackTrace();
    } catch (IOException e) {
        e.printStackTrace();
    }
}

public ArrayList<GameObject> readData() {
    ArrayList<GameObject> objectList = null;
    ObjectInputStream in;
    try {
        in = new ObjectInputStream(new FileInputStream("data/objectFileAll.obj"));
        objectList = (ArrayList<GameObject>) in.readObject();
        in.close();
    } catch (FileNotFoundException e) {
        e.printStackTrace();
    } catch (IOException e) {
        e.printStackTrace();
    } catch (ClassNotFoundException e) {
        e.printStackTrace();
    }
    return objectList;
}

@Override
public void initSprite() {
    final int SIZE = 8;
    screen = new Screen(SIZE);
    Random random = new Random();
    bullets = new HashSet<Plane.Bullet>();
    ArrayList<GameObject> objectList = readData();

    if(objectList == null) {                    // 如果初始化数据为空
        plane = new Plane(7,4);
    }
    else {
        for(GameObject object:objectList) {
            if(object instanceof Plane) {
                plane = (Plane) object;
            }
            if(object instanceof Plane.Bullet bullet) {
                bullets.add(bullet);
            }
        }
    }
    screen.add(bullets);
    screen.add(plane);
```

```java
        bullet = null;
    }

    @Override
    public void update() {
        char key = screen.getKey();
        screen.delay();
        if(key == 'w') {
            plane.moveUp();
        }
        if(key == 's') {
            plane.moveDown();
        }
        if(key == 'a') {
            plane.moveLeft();
        }
        if(key == 'd') {
            plane.moveRight();
        }
        if(key == 'k') {

            bullet = plane.shoot();
            screen.add(bullet);
            bullets.add(bullet);
        }

        if(key == 'p') {
            try {
                saveData(screen.getData());    //获得屏幕上的数据,并且将其保存
            } catch (Exception e) {
                e.printStackTrace();
            }
        }

        for(Plane.Bullet blet:bullets) {
            blet.moveUp();
        }
    }
}
```

测试代码如下:

```java
public class Main {
    public static void main(String[ ] args){
        PlaneGame game = new PlaneGame();
        game.init();
        game.run();
    }
}
```

运行程序,按下按键"p"对游戏进行存档。重启游戏后,游戏的初始界面就是存档时的界面。
为了减轻学习负担,本章案例中所使用的读写操作都是标准IO。Java语言中引入了增强版的

IO 功能，也称为 New IO API，简称 NIO。NIO 与标准 IO 的作用和目的相同，但是使用方式不同，NIO 采用内存映射文件的方式处理输入输出，使用的是通道和缓冲区，可以像访问内存一样访问文件。标准 IO 中，使用的是字节流和字符流。相比较标准 IO，NIO 以更高效的方式读写文件。

在 JDK 7 中，进一步对 NIO 进行了扩展，增强了对文件处理和文件系统特性的支持，提供了异步非堵塞 IO 操作方式，称为 NIO.2，读者如果对此感兴趣，可以自学相关内容。

习题

10.1 下面关于 FileInputStream 类型的说法中不正确的是_____。
   A. 创建 FileInputStream 对象是为了读取硬盘上的文件。
   B. 创建 FileInputStream 对象时，如果硬盘上对应的文件不存在，则抛出一个异常。
   C. FileInputStream 对象不可以创建文件。
   D. FileInputStream 对象读取文件时，只能读取文本文件。

10.2 使用 Java 输入输出流实现对文本文件的读写过程中，需要处理下列_____异常。
   A. ClassNotFoundException   B. IOException
   C. SQLException   D. RemoteException

10.3 下列流中_____使用了缓冲区技术。
   A. BufferedOutputStream   B. FileInputStream
   C. DataOutputStream   D. FileReader

10.4 编写程序，实现"贪吃蛇"游戏的存档和读档功能。

第11章

多 线 程

多线程编程是 Java 语言非常重要的特点之一。在程序设计中,多线程就是单个程序内可同时运行多个不同的线程来完成任务。比如,游戏程序内,同一时间既可以玩游戏,还能与队友聊天。采用多线程技术的程序充分利用了 CPU 的空闲时间,用尽可能少的时间来对用户的要求做出响应,使得进程的整体运行效率得到较大提高,同时增强了应用程序的灵活性。

11.1 线程的概念

线程的概念来源于进程。进程就是每个独立执行的程序,如正在运行的 QQ 聊天程序、Word 编写程序等。在多任务操作系统中,可以查看当前系统中所有的进程。每一个进程都有一组独立的系统资源。线程是进程中的实际运作单位,被包含在进程中,是操作系统能够进行运算调度的最小单位,也被称为轻量级进程。一个进程中可以并发多条线程,每条线程并行执行不同的任务。同一进程的所有线程共享同一内存和同一组系统资源,使得不同任务之间的协调操作与运行、数据的交互、资源的分配等问题更加易于解决。

与进程相比,线程是一种非常"节俭"的多任务操作方式。启动一个线程所花费的空间远远小于启动一个进程所花费的空间。而且,线程间相互切换所需的时间也远远小于进程间相互切换所需要的时间。与进程相比,线程间的通信机制更方便。对不同进程来说,它们具有独立的数据空间,要进行数据的传递只能通过通信的方式进行,这种方式不仅费时,而且很不方便。线程则不然,同一进程下的线程之间共享数据空间,一个线程的数据可以直接为其他线程所用,这不仅快捷,而且方便。

使用多线程技术,将耗时长的操作置于一个新的线程,使多 CPU 系统更加有效。许多程序都是采用多线程技术实现,比如游戏程序,界面的渲染、网络通信都是分别工作,采用不同的线程完成。Java 语言对多线程技术提供了非常完善的机制,让程序设计者能够高效地编写多线程程序。

11.2 线程的创建

在 Java 语言中,实现多线程的主要方式有如下两种:
(1) 继承 Thread 类,重写 run()方法。
(2) 实现 Runnable 接口,重写 run()方法。
其中,run()方法的方法体就是线程要完成的任务。

11.2.1 继承 Thread 类实现多线程

通过继承 Thread 类实现多线程的方法如下:
(1) 创建 Thread 类的子类,重写 run()方法。
(2) 创建该子类的实例对象,并调用线程对象的 start()方法启动线程。

【例 11.1】 通过继承 Thread 类实现多线程。

设计一个线程类 TreadDemo,继承于 Thread 类,主要完成输出 5 次线程名,代码如下:

视频讲解

```
public class ThreadDemo extends Thread{
    public ThreadDemo(String name) {
        super(name);
    }
    public void run() {
        for(int i = 0; i < 5; i++) {
            System.out.println(getName() + " running");
        }
    }
}
```

开启两个线程,测试输出结果,测试代码如下:

```
public class Main {
    public static void main(String[ ] args) throws Exception{
        ThreadDemo threadA = new ThreadDemo("A");
        threadA.start();
        ThreadDemo threadB = new ThreadDemo("B");
        threadB.start();
    }
}
```

程序运行结果如图 11.1 所示。

通过运行结果可知,两个线程实例对象交替运行,而不是按照代码顺序,执行完第一个线程方法后,再执行第二个线程方法,这就是多线程的效果,即多段程序交替运行。

11.2.2 通过 Runnable 接口实现多线程

通过继承 Thread 类的方式可以实现多线程,但是这种方法有时候会遇到困境。比如,一个类

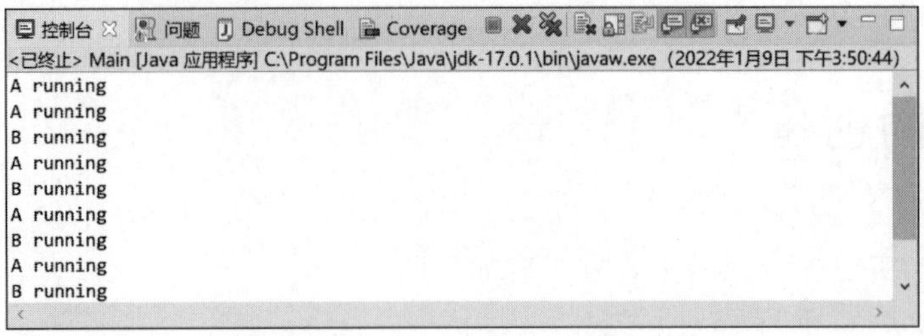

图 11.1 通过继承 Thread 类实验多线程示例运行结果

已经继承了其他父类,就无法通过继承 Thread 类实现多线程,因为 Java 语言只支持单继承。遇到这种情况,可以通过实现 Runnable 接口的方式实现多线程。

通过 Runnable 接口实现多线程的方法如下:

(1) 创建 Runnable 接口的实现类,并重写该接口的 run()方法。

(2) 创建 Runnable 实现类的对象,并将此对象作为 Thread 的参数来创建 Thread 对象,该 Thread 对象才是真正的线程对象。

(3) 调用线程对象的 start()方法启动该线程。

【例 11.2】 通过 Runnable 接口实现多线程。

设计一个线程类 RunnableDemo,通过 Runnable 实现多线程,代码如下:

```java
public class RunnableDemo implements Runnable{
    public void run() {
        for(int i = 0; i < 5; i++) {
            System.out.println(Thread.currentThread().getName() + " running");
        }
    }
}
```

开启两个线程,测试运行结果,测试代码如下:

```java
public class Main {
    public static void main(String[ ] args) throws Exception{
        RunnableDemo runnable = new RunnableDemo();
        Thread threadA = new Thread(runnable,"A");
        threadA.start();
        Thread threadB = new Thread(runnable,"B");
        threadB.start();
    }
}
```

程序运行结果如图 11.2 所示。

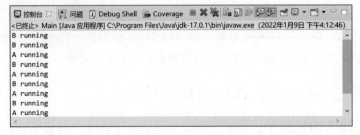

图 11.2 通过 Runnable 接口实现多线程示例运行结果

通过运行结果可知,两个线程实例对象交替运行。通过例子可知,通过实现 Runnable 接口的方式也可以实现多线程。

11.3 线程的状态与调度

多线程程序中,经常会遇到这样的场景:一个线程运行到一半时,新的线程需要立即执行。如何有效对线程进行管理,离不开线程的状态和调度。

11.3.1 线程的状态

每个事物都有其生命周期,也就是事物从出生到消亡的过程。线程也不例外,也有自己的生命周期。在线程的整个生命周期中存在 6 个状态,在某一时刻线程只能是这 6 种状态中的一种。这 6 种状态具体如下。

1. 新建状态

创建一个线程对象后,该线程对象就处于新建状态(New),此时它不能运行,与其他 Java 对象一样,仅仅由 Java 虚拟机为其分配了内存,没有表现出任何线程的动态特征。

2. 可运行状态

可运行状态(Runnable)对应操作系统线程状态中的两种状态,分别是就绪(Ready)和运行(Running)。当线程对象调用了 start()方法后,该线程就进入就绪状态。处于就绪状态的线程位于线程队列中,此时它只是具备了运行的条件。如果处于就绪状态的线程获得了 CPU 的使用权,并开始执行 run()方法中的线程执行体,则该线程处于运行状态。一个线程启动后,它可能不会一直处于运行状态,当运行状态的线程使用完系统分配的时间后,系统就会剥夺该线程占用的 CPU 资源,让其他线程获得执行的机会。需要注意的是,只有处于就绪状态的线程才可能转换到运行状态。

3. 阻塞状态

一个正在执行的线程在某些特殊情况下,如被人为挂起或执行耗时的输入输出操作时,会让出 CPU 的使用权并暂时中止执行,进入阻塞状态(Blocked)。线程进入阻塞状态后,就不能进入排队队列。只有当引起阻塞的原因被消除后,线程才可以转入就绪状态。

4. 等待状态

当处于运行状态下的线程调用 Thread 类的 wait()方法时,该线程就会进入等待状态(Waiting)。进入等待状态的线程必须调用 Thread 类的 notify()方法或者 notifyAll()方法才能被唤醒。

5. 计时等待状态

计时等待状态(Timed Waiting)与等待状态非常相似,其中的区别只在于等待是否有时间的限制。计时等待状态设置了等待时间,超时会由系统唤醒,其间也可以提前被唤醒,如使用 notify()方法或者 notifyAll()方法将其唤醒。

6. 结束状态

如果线程调用 stop()方法或者 run()方法正常执行完毕,或者线程抛出一个未捕获的异常错误,线程就进入结束状态(Terminated)。一旦进入结束状态,线程将不再拥有运行的资格,也不能再转换到其他状态。

线程的不同状态表明了线程当前正在进行的活动,在程序中,通过一些操作,可以使线程的状态发生转变,如图 11.3 所示。

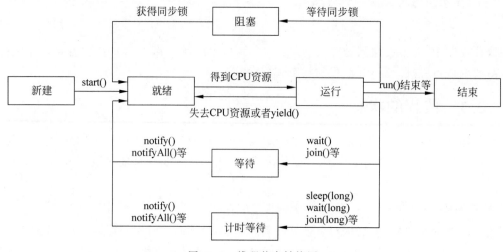

图 11.3　线程状态转换图

11.3.2　线程的调度

1. 线程的优先级

程序中的多个线程可以并发运行,但是并不在同一时刻运行。对于单 CPU 的计算机,一个时刻只能有一个线程处于运行状态。多线程的并发运行,实际上是各个线程轮流获得 CPU 的使用权,分别执行各自的任务。在运行池中,会有多个处于就绪状态的线程在等待 CPU,Java 虚拟机的一项非常重要的任务就是负责线程的调度,按照特定机制为多个线程分配 CPU 的使用权。

Java 虚拟机默认采用抢占式调度模型,每个线程都有一个优先级,当多个线程处于可运行状态,线程调度器优先让运行池中优先级高的线程占用 CPU。如果运行池中的线程优先级相同,那么就随机选择一个线程,使其占用 CPU。

线程的优先级用 1~10 的整数表示,数值越大优先级越高。线程的优先级可以通过 Thread 类的 setPriority()方法进行设置。

【例 11.3】　设置线程的优先级。

设计两个线程 A 和 B,优先级分别是 1 和 10,实现同一操作,即将 0~999 依次打印出来,代码如下:

视频讲解

```java
public class RunnableDemo implements Runnable{
    public void run() {
        for(int i = 0; i < 1000; i ++ ) {
            System.out.println(Thread.currentThread().getName() + " running");
        }
    }
}
```

测试代码如下：

```
public class Main {
    public static void main(String[ ] args) throws Exception{
        RunnableDemo runnable = new RunnableDemo();
        Thread threadA = new Thread(runnable,"A");
        Thread threadB = new Thread(runnable,"B");
        threadA.setPriority(1);
        threadB.setPriority(10);
        threadA.start();
        threadB.start();
    }
}
```

运行程序，线程 B 比线程 A 先结束，因为它的优先级更高，会占用更多的 CPU 使用权。

2. join()方法

join()方法的作用是阻塞指定的线程，直到另一个线程完成以后再继续执行，相当于线程被另外一个线程插队。

【例 11.4】 设置线程的阻塞。

设计两个线程：主线程和线程 A，任务都是将 0～4 打印出来，主线程打印完 2 之后，线程 A 插队，优先执行。使用 join()方法进行插队时，需要注意 join()方法声明了抛出异常，因此在调用该方法时应捕获异常，或声明抛出该异常，代码如下：

视频讲解

```
public class RunnableDemo implements Runnable{
    public void run() {
        for(int i = 0; i < 5; i++) {
            System.out.println(Thread.currentThread().getName() + " running");
        }
    }
}
```

测试代码如下：

```
public class Main {
    public static void main(String[ ] args) throws Exception{
        RunnableDemo runnable = new RunnableDemo();
        Thread thread = new Thread(runnable,"A");

        for(int i = 0;i < 5;i++) {
            System.out.println("main " + i);
            if(i == 2) {
                thread.start();
                thread.join();
            }
        }
    }
}
```

程序运行结果如图 11.4 所示。

通过结果可知，线程 A 插队之后，当前 main 线程会被挂起，让线程 A 执行，直到线程 A 执行

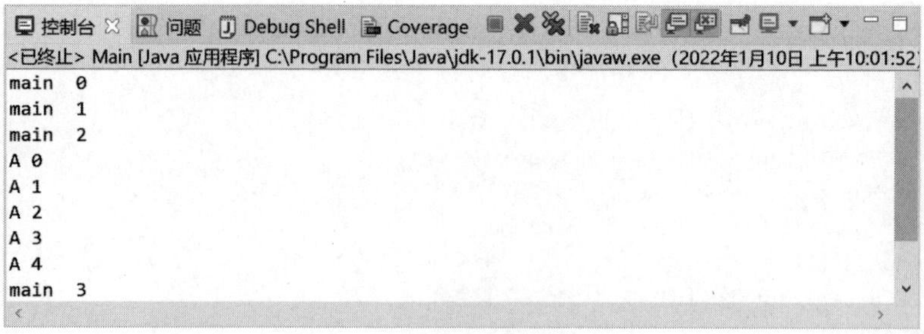

图 11.4　线程阻塞示例运行结果

完之后,main 线程才会重新被执行。

3. sleep()方法

sleep()方法是让当前线程暂停一段时间,进入休眠等待状态,这个时候其他正在等待被执行的线程将有机会被执行。

【例 11.5】　设置线程的休眠。

sleep()方法声明了抛出异常,因此在调用该方法时应捕获异常,或声明抛出该异常。设计两个线程 A 和 B,线程 A 打印 0～9,线程 B 打印 0～4。线程 A 的代码如下:

```java
public class ThreadDemoA extends Thread{
    public ThreadDemoA(String name) {
        super(name);
    }

    public void run() {
        for(int i = 0; i < 10; i++) {
            System.out.println(getName() + " " + i);
        }
    }
}
```

线程 B 的代码如下:

```java
public class ThreadDemoB extends Thread{
    public ThreadDemoB(String name) {
        super(name);
    }

    public void run() {
        for(int i = 0; i < 5; i++) {
            System.out.println(getName() + " " + i);
            try {
                Thread.sleep(1000);                    //休眠一秒
            } catch (Exception e) {
            }
        }
    }
}
```

测试代码如下:

```
public class Main {
    public static void main(String[ ] args) throws Exception{
        ThreadDemoA threadA = new ThreadDemoA("A");
        ThreadDemoB threadB = new ThreadDemoB("B");
        threadA.start();
        threadB.start();
    }
}
```

程序运行结果如图 11.5 所示。

图 11.5　线程休眠示例的运行结果

创建了两个线程 threadA 和 threadB,它们的优先级相同,虽然线程 A 打印 10 个数,线程 B 只打印 5 个数,但是线程 threadB 执行过程中调用了 sleep()方法,目的是让线程在某一个时刻休眠一段时间,从而使线程 threadA 获得 CPU 使用权,所以会优先打印完数据。

4. yield()方法

yield()方法的作用是放弃当前线程获得的执行权,让其他线程有机会获得执行权。需要注意的是当前线程仍然可以和其他等待执行的线程一起竞争处理器资源。

【例 11.6】　设置线程让步。

设计两个线程 A 和 B,任务都是打印 0~9,线程 A 的代码如下:

```
public class ThreadDemoA extends Thread{
    public ThreadDemoA(String name) {
        super(name);
    }

    public void run() {
        for(int i = 0; i < 10; i ++ ) {
            System.out.println(getName() + " " + i);
        }
    }
}
```

线程 B 的代码如下：

```java
public class ThreadDemoB extends Thread{
    public ThreadDemoB(String name) {
        super(name);
    }
    public void run() {
        for(int i = 0; i < 10; i ++ ) {
            System.out.println(getName() + " " + i);
            Thread.yield();
        }
    }
}
```

测试代码如下：

```java
public class Main {
    public static void main(String[ ] args) throws Exception{
        ThreadDemoA threadA = new ThreadDemoA("A");
        ThreadDemoB threadB = new ThreadDemoB("B");
        threadA.start();
        threadB.start();
    }
}
```

程序运行结果如图 11.6 所示。

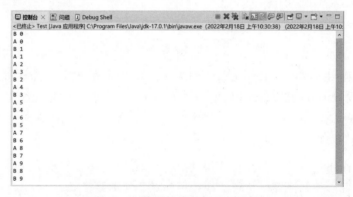

图 11.6　线程让步示例运行结果

通过结果可知，yield()方法与 sleep()方法的区别是：yield()方法不会阻塞该线程，只是将线程转换成就绪状态，让当前正在运行的线程失去 CPU 使用权，使系统调度器重新调度一次。所有线程，包括当前线程，会再次抢占 CPU 资源使用权。

11.4　线程同步与对象锁

11.4.1　线程安全

多线程技术可以更好地利用系统资源，但是也会带来问题：当多个线程去访问同一个资源时，会引发线程安全问题。例如,经典的"银行取款"问题,账户上的余额有 1000 元,同时从银行柜台和取款机上取出 800 元,正常情况下,柜台和取款机两者中只有一个能成功。但是,如果采用多线程模拟同一账户并发取钱操作,有可能导致都能成功。

【例 11.7】 模拟从银行取钱。

设计一个银行账号类,并且设计线程类模拟柜台和取款机取款,银行账户类代码如下:

```java
public class Account {
    private int balance = 1000;              //账户余额
    public int getBalance() {
        return balance;
    }

    public boolean withdraw (int num) {      //取钱
        if ( balance >= num) {               //如果余额大于或等于要取钱的数目
            try {
                Thread.sleep(1000);
            } catch (InterruptedException e) {
                e.printStackTrace();
            }

            balance -= num;                  //扣除所取得钱
            return true;                     //取钱成功
        }
        else {
            return false;
        }
    }
}
```

取钱线程类代码如下:

```java
public class ThreadBank extends Thread{
    Account account;
    int num;
    ThreadBank(Account account, int num) {
        this.account = account;
        this.num = num;
    }

    @Override
    public void run() {
        if(account.withdraw (num)){
            System.out.println(account.getBalance());
        }else{
            System.out.println("余额不足,不能取款");
        }
    }
}
```

测试代码如下:

```java
public class Main {
    public static void main(String[ ] args) throws Exception{
        Account account =  new Account();
        ThreadBank threadCounter = new ThreadBank(account,800);      //柜台取 800 元
        ThreadBank threadATM = new ThreadBank(account,800);          //ATM 取 800 元
        threadCounter.start();
        threadATM.start();
    }
}
```

运行程序,结果可能出现的情况如图 11.7 所示。

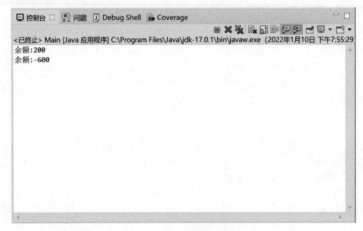

图 11.7　模拟银行示例

通过运行结果可知,两次取款都成功了,导致账户出现了-600 元情况,产生的原因是由于多线程操作共享资源导致了线程安全问题。为了解决这样的问题,需要实现多线程的同步,限制某个资源在同一时刻只能被一个线程访问。

11.4.2　同步方法

Java 程序的每一个对象都有一个内置锁,是通过 synchronized 关键字实现的。用 synchronized 关键字修饰的方法被称为同步方法。线程访问某个对象的同步方法时,将该对象上锁,此时其他任何线程都无法再访问该对象的同步方法了,直到当前线程执行方法完毕后(或者抛出了异常),那么将该对象的锁释放掉,其他线程才有可能再去访问该同步方法。

视频讲解

【例 11.8】　使用同步方法模拟银行取钱。

在 Account 类的 withdraw()方法前面加上 synchronized 关键字,使之成为同步方法,保障数据安全,代码如下:

```java
public class Account {
    private int balance = 1000;
    public int getBalance() {
        return balance;
    }
```

```java
public synchronized boolean withdraw (int num) {        //取钱
    if ( balance > num) {
        try {
            Thread.sleep(1000);
        } catch (InterruptedException e) {
            e.printStackTrace();
        }
        balance -= num;
        return true;
    }
    else {
        return false;
    }
}
```

取钱线程类代码如下：

```java
public class ThreadBank extends Thread{
    Account account;
    int num;
    ThreadBank(Account account,int num) {
        this.account = account;
        this.num = num;
    }

    @Override
    public void run() {
        if(account.withdraw(num)){
            System.out.println(account.getBalance());
        }else{
            System.out.println("余额不足,不能取款");
        }
    }
}
```

测试代码如下：

```java
public class Main {
    public static void main(String[ ] args) throws Exception{
        Account account = new Account();
        ThreadBank threadCounter = new ThreadBank(account,800);
        ThreadBank threadATM = new ThreadBank(account,800);
        threadCounter.start();
        threadATM.start();
    }
}
```

程序运行结果如图 11.8 所示。

使用同步方法后,就保证了不会出现余额为负数的情况。

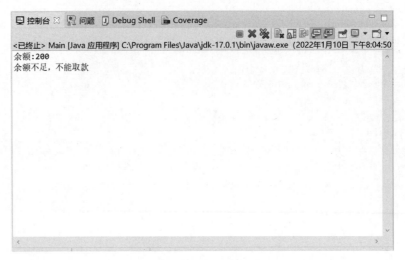

图 11.8 同步方法示例结果

11.4.3 同步代码块

同步是一种非常耗时的操作,因此应该尽量减少同步的内容。通常没有必要同步整个方法,使用 synchronized 关键字同步关键代码块即可。同步代码块的格式如下:

```
synchronized(同步监视器){
    //需要同步的代码块
}
```

同步监视器是一个锁对象。同步代码块是通过锁定一个指定的对象来对同步块中包含的代码进行同步。

视频讲解

【例 11.9】 使用同步代码块模拟银行取钱。

如果 Account 类的 withdraw()方法没有同步,可以在线程类中同步代码块,代码如下:

```java
public class Account {
    private int balance = 1000;
    public int getBalance() {
        return balance;
    }

    public boolean withdraw(int num) {

        if ( balance >= num) {
            try {
                Thread.sleep(1000);
            } catch (InterruptedException e) {
                e.printStackTrace();
            }

            balance -= num;
            return true;
        }
```

```
            else {
                return false;
            }
        }
    }
```

取钱线程类代码如下:

```
public class ThreadBank extends Thread{
    Account account;
    int num;
    ThreadBank(Account account, int num) {
        this.account = account;
        this.num = num;
    }

    @Override
    public void run() {
        synchronized(account) {
            if(account.withdraw(num)){
                System.out.println(account.getBalance());
            }else{
                System.out.println("余额不足,不能取款");
            }
        }
    }
}
```

测试代码如下:

```
public class Main {
    public static void main(String[ ] args) throws Exception{
        Account account = new Account();
        ThreadBank threadCounter = new ThreadBank(account,800);
        ThreadBank threadATM = new ThreadBank(account,800);
        threadCounter.start();
        threadATM.start();
    }
}
```

程序运行结果如图 11.9 所示。

使用同步代码块,当一个线程要访问 Account 对象时,必须获得对象上的锁,直到同步代码块执行结束后才释放对象锁。

11.4.4　同步锁

使用 synchronized 关键字同步方法或者代码块实现保护线程安全的方法,使用方式简单,但是也有一些不足的地方。例如,如果一个线程获得了锁之后,进入了阻塞状态,但是锁不会被释放,其他线程就需要一直等待才有机会获得锁,这样会降低执行效率。所以就需要一个在线程阻塞时可以释放线程锁的新方案,Java 语言提供了功能更强大的同步 Lock 锁来解决这个问题。

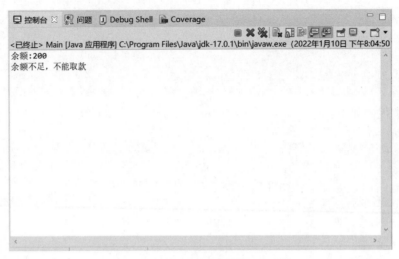

图 11.9　同步代码块示例的运行结果

Lock 锁的优势在于可以让某个线程在持续获取同步锁失败后返回,不再继续等待,这样就提高了效率。

Lock 锁的使用方法是通过 Lock 接口的实现类来创建一个 Lock 锁对象,并通过 Lock 锁对象的 lock()方法和 unlock()方法对核心代码进行上锁和解锁。

视频讲解

【例 11.10】　使用同步锁模拟银行取钱。

使用 Lock 锁模拟银行取钱过程的代码如下:

```java
public class Account {
    private int balance = 1000;
    Lock lock = new ReentrantLock();
    public int getBalance() {
        return balance;
    }

    public boolean withdraw(int num) {
        lock.lock();
        if ( balance > num) {
            try {
                Thread.sleep(1000);
            } catch (InterruptedException e) {
                e.printStackTrace();
            }
            balance -= num;
            lock.unlock();
            return true;
        }
        else {
            return false;
        }
    }
}
```

取钱线程类代码如下:

```java
public class ThreadBank extends Thread{
    Account account;
    int num;
    ThreadBank(Account account,int num) {
        this.account = account;
        this.num = num;
    }

    @Override
    public void run() {
        if(account.withdraw(num)){
            System.out.println(account.getBalance());
        }else{
            System.out.println("余额不足,不能取款");
        }
    }
}
```

测试代码如下：

```java
public class Main {
    public static void main(String[ ] args) throws Exception{
        Account account = new Account();
        ThreadBank threadCounter = new ThreadBank(account,800);
        ThreadBank threadATM = new ThreadBank(account,800);
        threadCounter.start();
        threadATM.start();
    }
}
```

程序运行结果如图 11.10 所示。

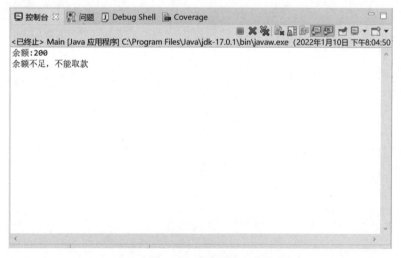

图 11.10　同步锁示例的运行结果

通过 Lock 接口的实现类 ReentrantLock 来创建一个 Lock 锁对象,并通过 Lock 锁对象的 lock()方法和 unlock()方法对核心代码进行上锁和解锁,通过结果可以看出,使用同步锁也可以正常取钱,解决了线程安全问题。

11.4.5 死锁问题

线程同步保证了数据安全,但是有时也会导致死锁问题,当线程需要同时持有多个锁时,有可能产生死锁。例如,线程 A 当前持有同步锁 lock1,线程 B 当前持有同步锁 lock2。当线程 A 持有 lock1 的同时试图获取 lock2 时,由于线程 B 正持有 lock2,因此线程 A 会阻塞等待线程 B 释放 lock2。如果此时线程 B 在持有 lock2 的同时,也试图获取 lock1,由于线程 A 正持有 lock1,因此线程 B 会阻塞等待 A 对 lock1 的释放。此时二者都在等待对方释放所持有的锁,而二者却又都没释放自己所持有的锁,这时二者便会一直阻塞下去,这种情形称为死锁。

在计算机系统中也存在类似的情况。例如,某计算机系统中只有一台打印机和一台输入设备,进程 P1 正占用输入设备,同时又提出使用打印机的请求,但此时打印机正被进程 P2 所占用,而 P2 在未释放打印机之前,又提出请求使用正被 P1 占用着的输入设备。这样两个进程便会一直等待下去,均无法继续执行,陷入死锁状态。

视频讲解

【例 11.11】 模拟银行账户转账出现死锁的情况。

两个银行账户 A 和 B 相互转账的时候,当账户 A 正向账户 B 转账的时候,账户 B 也正向账户 A 转账,可能会产生死锁情况出现,代码如下:

```java
public class Account {
    private String name;
    private int balance = 1000;

    Account(String name){
        this.name = name;
    }
    public int getBalance() {
        return balance;
    }

    public void withdraw(int num) {                    //取钱
        try {
            Thread.sleep(1000);
        } catch (InterruptedException e) {
            e.printStackTrace();
        }

        if ( balance > num) {
            balance -= num;
            System.out.println(name + "余额:" + balance);
        }
        else {
            System.out.println("余额不足,不能取款");
        }
    }

    public void deposit(int num) {                     //存钱
        try {
```

```java
            Thread.sleep(1000);
        } catch (InterruptedException e) {
            e.printStackTrace();
        }
        balance + = num;
        System.out.println(name + "余额:" + balance);
    }
}
```

转账线程类代码如下：

```java
public class ThreadBank extends Thread{
    Account fromAccount;
    Account toAccount;
    int num;

    public ThreadBank(Account fromAccount, Account toAccount,int num) {
        this.fromAccount = fromAccount;
        this.toAccount = toAccount;
        this.num = num;
    }

    public static boolean transfer(Account fromAccount,Account toAccount,int amount){    //转账
        synchronized (fromAccount) {                       //锁定汇款账户
            try {
                Thread.sleep(500);
            } catch (Exception e) {
                e.printStackTrace();
            }
            synchronized (toAccount) {                     //锁定来款账户
                if (fromAccount.getBalance() < amount) {   //判断余额是否大于0
                    return false;
                } else {
                    fromAccount.withdraw(amount);          //汇款账户减钱
                    toAccount.deposit(amount);             //来款账户增钱
                    return true;
                }
            }
        }
    }

    @Override
    public void run() {
        if(transfer(fromAccount,toAccount,num)) {
            System.out.println("转账成功");
        }
        else {
            System.out.println("转账失败");
        }
    }
}
```

测试代码如下:

```java
public class Main {
    public static void main(String[ ] args) throws Exception{
        Account accountA = new Account("A");
        Account accountB = new Account("B");

        ThreadBank threadA = new ThreadBank(accountA,accountB,800);
        ThreadBank threadB = new ThreadBank(accountB,accountA,200);
        threadA.start();
        threadB.start();
    }
}
```

运行程序,没有任何结果,因为发生死锁。线程 A 和线程 B 都在等待对方所持有锁的释放,而二者却又都没释放自己所持有的锁,于是两个线程都处于挂起状态,从而导致了线程死锁。

避免死锁的方法如下:

(1) 设计时尽量避免嵌套锁的情况,每次只锁定一个对象。如果无法避免嵌套锁的情况,按照一定的顺序加锁。

(2) 设置加锁时限,线程获取锁的过程中限制一定的时间,如果给定时间内获取不到,就放弃对锁的请求,并释放自己占有的锁。

【例 11.12】 解决模拟银行汇款出现死锁的情况。

需要给两个对象同时加锁时,可以按照一定的顺序进行加锁。比如,根据对象的 hashcode 值大小来确定先后顺序,也可以给账号增加一个属性,如 id 值,根据 id 值大小确定锁的顺序,代码如下:

```java
public class Account {
    private int id;
    private String name;
    private int balance = 1000;

    Account(String name, int id){
        this.name = name;
        this.id = id;
    }
    public int getBalance() {
        return balance;
    }

    public int getID(){
        return id;
    }
    public void withdraw(int num) {
        try {
            Thread.sleep(1000);
        } catch (InterruptedException e) {
            e.printStackTrace();
        }

        if ( balance > num) {
            balance -= num;
```

```java
            System.out.println(name + "余额:" + balance);
        }
        else {
            System.out.println("余额不足,不能取款");;
        }
    }

    public void deposit (int num) {
        try {
            Thread.sleep(1000);
        } catch (InterruptedException e) {
            e.printStackTrace();
        }
        balance + = num;
        System.out.println(name + "余额:" + balance);
    }
}
```

转账线程类代码如下：

```java
public class ThreadBank extends Thread{
    Account fromAccount;
    Account toAccount;
    int num;

    public ThreadBank(Account fromAccount, Account toAccount,int num) {
        this.fromAccount = fromAccount;
        this.toAccount = toAccount;
        this.num = num;
    }

    public static boolean transfer(Account fromAccount,Account toAccount,int amount){    //转账
        if(fromAccount.getID() < toAccount.getID()) {
            synchronized (fromAccount) {                          //锁定汇款账户
                try {
                    Thread.sleep(500);
                } catch (Exception e) {
                    e.printStackTrace();
                }
                synchronized (toAccount) {                        //锁定来款账户
                    if (fromAccount.getBalance() < amount) {      //判断余额是否大于0
                        return false;
                    } else {
                        fromAccount.withdraw(amount);              //汇款账户减钱
                        toAccount.deposit(amount);                 //来款账户增钱
                        return true;
                    }
                }
            }
        }

        if(fromAccount.getID() >= toAccount.getID()) {
```

```java
            synchronized (toAccount) {                    //锁定汇款账户
                try {
                    Thread.sleep(500);
                } catch (Exception e) {
                    e.printStackTrace();
                }
                synchronized (fromAccount) {               //锁定来款账户
                    if (fromAccount.getBalance() < amount) {  //判断余额是否大于0
                        return false;
                    } else {
                        fromAccount.withdraw(amount);      //汇款账户减钱
                        toAccount.deposit(amount);         //来款账户增钱
                        return true;
                    }
                }
            }
        }
        return false;
    }

    @Override
    public void run() {
        if(transfer(fromAccount,toAccount,num)) {
            System.out.println(fromAccount.getID() + "转账成功");
        }
        else {
            System.out.println(fromAccount.getID() + "转账失败");
        }
    }
}
```

测试代码如下：

```java
public class Main {
    public static void main(String[ ] args) throws Exception{
        Account accountA = new Account("A",10001);
        Account accountB = new Account("B",10002);

        ThreadBank threadA = new ThreadBank(accountA,accountB,800);
        ThreadBank threadB = new ThreadBank(accountB,accountA,200);
        threadA.start();
        threadB.start();
    }
}
```

运行程序，转账成功。

通过一定的顺序来固定加锁的顺序，这样就不会发生死锁的情况了。

11.4.6 线程通信

一般而言，在一个进程中，线程往往不是孤立存在的，常常需要和其他线程通信，以执行特定的任务。例如，主线程和次线程、次线程与次线程、工作线程和用户界面线程之间的通信等。这样,线程与线程之间必定有一个信息传递的渠道。

为了更好地理解线程间的通信,可以模拟现实生活中常见的生产者与消费者模式。这个模式中,一部分线程用于生产数据,另一部分线程用于处理数据,于是便有了形象的生产者与消费者模式。举一个简单的例子,有一个生产者生产包子,他将生产好的包子放到筐中,由消费者从筐中拿走包子。

在生产的过程会出现以下两种情况:生产者的速度比消费者快,那么消费者还没来得及将包子拿走,生产者又生产了新的包子,长此以往,装包子的筐就装满了,不能继续生产了。反之,消费者的速度比生产者快,消费者就需要一直等待,直到有新的包子产生为止。

生产者与消费者模式应用场景非常广泛,在程序设计中经常会遇到。例如,有一个数据缓冲区,一个或者多个生产者把数据存入这个缓冲区,一个或者多个消费者取出数据缓冲区中的共享数据,当缓冲区满的时候生产者将不能再放入数据到缓冲区(生产者线程阻塞),当缓冲区空的时候消费者不能从缓冲区中读取数据(消费者线程阻塞)。

【例 11.13】 模拟银行信用卡业务,可以透支银行卡,但是透支额度不超过 300 元。

这是一个经典的生成者与消费者模型,一边使用信用卡提前消费,一边往信用卡里存钱还贷,代码如下:

视频讲解

```java
public class Account {
    private String name;
        private int balance = 0;

        Account(String name){
            this.name = name;
        }
        public synchronized int getBalance() {
            return balance;
        }

        public synchronized void withdraw(int num) {           //取钱或者借贷
            try {
                Thread.sleep(1000);
            } catch (InterruptedException e) {
                e.printStackTrace();
            }
            balance -= num;
            System.out.println("消费" + num +"元成功," + name + "余额:" + balance);
        }

        public synchronized void deposit(int num) {            //存钱或者还贷
            try {
                Thread.sleep(1000);
            } catch (InterruptedException e) {
                e.printStackTrace();
            }
            balance += num;
            System.out.println("还款" + num +"元成功,"+ name + "余额:" + balance);
        }
}
```

还贷线程类代码如下:

```java
public class ThreadDeposite extends Thread{
    Account account;
    public ThreadDeposite (Account account) {
        this.account = account;
    }

    @Override
    public void run() {
        for(int i = 0;i<10;i++){
            account.deposit(100);
        }
    }
}
```

借贷线程类代码如下:

```java
public class ThreadWithdraw extends Thread{
    Account account;
    public ThreadWithdraw (Account account) {
        this.account = account;
    }

    @Override
    public void run() {
        for(int i = 0;i<10;i++){
            account.withdraw(100);
        }
    }
}
```

测试代码如下:

```java
public class Main {
    public static void main(String[ ] args) throws Exception{
        Account account = new Account("A");

        ThreadWithdraw threadA = new ThreadWithdraw(account);
        ThreadDeposite threadB = new ThreadDeposite(account);
        threadA.start();
        threadB.start();
    }
}
```

程序运行结果如图 11.11 所示。

从图中结果可知,出现了透支额度超过 300 元的情况。为了实现确保透支额度不超过 300 元的情况,则需要在两个线程之间进行通信,保证线程间任务协调进行。Java 语言提供了 wait()、notify()、notifyAll()等方法用于线程间的通信。

(1) wait()方法。使当前线程放弃同步锁并进入等待,直到其他线程进入此同步锁。

(2) notify()方法。唤醒此同步锁上等待的第一个调用 wait()方法的线程。

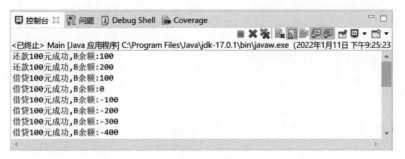

图 11.11 模拟信用卡业务的通行结果

（3）notifyAll()方法。唤醒此同步锁上所有调用 wait()方法的线程。

wait()方法用于使当前线程进入等待状态，notify()和 notifyAll()方法用于唤醒当前处于等待状态的线程。通过调用 wait()方法、notify()和 notifyAll()方法实现线程间的通信。

修改后的代码如下：

```java
public class Account {
    private String name;
    private int balance = 0;

    Account(String name){
        this.name = name;
    }
    public synchronized int getBalance() {
        return balance;
    }

    public synchronized void withdraw(int num) {
        try {
            Thread.sleep(500);
        } catch (InterruptedException e) {
            e.printStackTrace();
        }

        if(this.balance <= -300){
            try{
                wait();
            }catch(InterruptedException e){
            }
        }
        balance -= num;
        System.out.println("消费" + num + "元成功," + name + "余额:" + balance);
        notify();
    }

    public synchronized void deposit(int num) {
        try {
            Thread.sleep(1000);
        } catch (InterruptedException e) {
            e.printStackTrace();
        }
```

```
        if(this.balance >= 0){
            try{
                wait();
            }catch(InterruptedException e){
            }
        }
        balance += num;
        System.out.println("还款" + num + "元成功," + name + "余额:" + balance);
        notify();
    }
}
```

程序运行结果如图 11.12 所示。

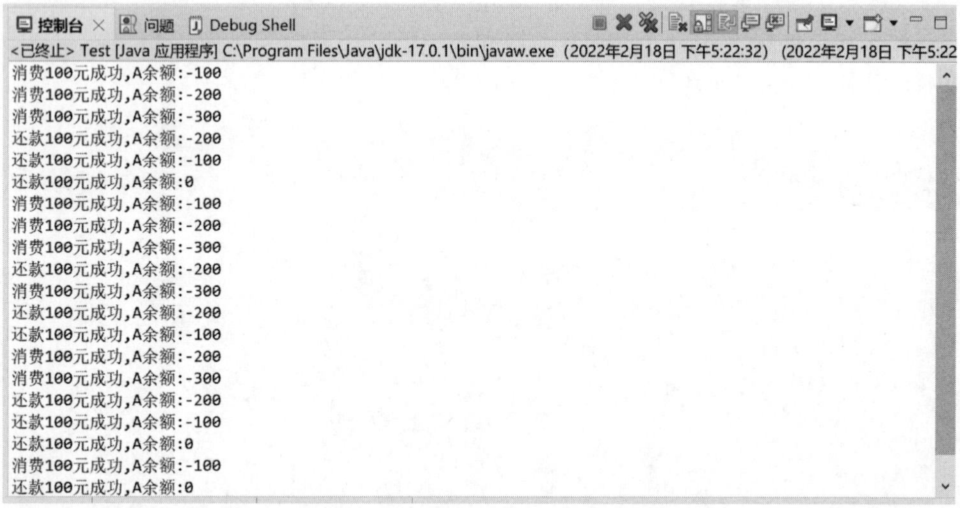

图 11.12 线程通信示例的运行结果

从图中结果可知,信用卡透支的额度没有超过 300 元的情况。

视频讲解

11.5 综合案例:多线程技术重构"飞机大战"游戏

飞机大战是一款经典的射击类游戏,游戏玩法非常简单,用户只需要控制自己的飞机上、下、左、右运动,并发送炮弹击落敌方飞机即可,如图 11.13 所示。

为了让游戏体验感更好,子弹在发射之后,运行速度越来越快。使用多线程技术可以控制子弹的速度,将子弹类设计成线程类,代码如下:

```
public class Plane extends GameObject implements Move{
    public Plane(int r, int c) {
        super(r,c);
    }

    @Override
    public void moveUp() {
```

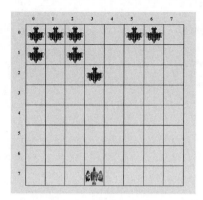

图 11.13 "飞机大战"游戏示意图

```
        if(getRow() > 0) {
            setRow(getRow() - 1);
        }
    }

    @Override
    public void moveDown() {
        if(getRow() + 1 < Config.SCREENSIZE) {
            setRow(getRow() + 1);
        }
    }

    @Override
    public void moveLeft() {
        if(getCol() > 0) {
            setCol(getCol() - 1);
        }
    }

    @Override
    public void moveRight() {
        if(getCol() + 1 < Config.SCREENSIZE) {
            setCol(getCol() + 1);
        }
    }

    public Bullet shoot() {
        Bullet bullet = new Bullet(getRow(),getCol());
        Thread thread = new Thread(bullet);
        thread.start();
        return bullet;
    }

public class Bullet extends GameObject implements Runnable{
    private int speed = 1;                    //设置速度
    private boolean state ;                   //设置子弹的状态
    public Bullet(int r, int c) {
        super(r,c);
```

```java
        }
        public void setSpeed(int speed) {
            this.speed = speed;
        }

        public int getSpeed() {
            return speed;
        }

        public void moveUp() {
            setRow(getRow() - 1);
        }

        public void run() {
            state = getLife();                    //获得子弹的生命状态
            while(state) {
                moveUp();
                if(getRow() < 0) {                //飞出了屏幕外
                    setLife(false);               //子弹的生命状态为消亡
                }
                try {
                    Thread.sleep(500 / speed);    //线程等待时间
                }catch(Exception e) {
                }
                if(speed < 5) {                   //子弹速度越来越快
                    speed ++ ;
                }
            }
        }
    }
}
```

其余代码不用修改,运行程序,按下按键"k"的时候,飞机发射子弹,并且发射的子弹运行速度越来越快。实现了控制子弹运行速度之后,接着实现子弹击中敌机后,敌机消失的功能。

设计一个敌机 Enemy 类,代码如下:

```java
public class Enemy extends GameObject {
    public Enemy(int r, int c) {
        super(r,c);
    }
}
```

设计好 Enemy 类之后,PlaneGame 类中也需要新增关于敌机的内容,为了避免有重复位置的敌机出现,使用 HashSet 集合存储敌机的信息,代码如下:

```java
public class PlaneGame extends GameCore{
    private Screen screen;
    Plane plane;
    HashSet<Plane.Bullet> bullets = new HashSet<Plane.Bullet>();
    HashSet<Enemy> enemys = new HashSet<Enemy>();
```

```java
    Plane.Bullet bullet;

    public void saveData(ArrayList<GameObject> objectList) {
        ObjectOutputStream out;
        try {
            out = new ObjectOutputStream(new FileOutputStream("data/objectFile.obj"));
            out.writeObject(objectList);
            out.close();
        } catch (FileNotFoundException e) {
            e.printStackTrace();
        } catch (IOException e) {
            e.printStackTrace();
        }
    }

    public ArrayList<GameObject> readData() {
        ArrayList<GameObject> objectList = null;
        ObjectInputStream in;
        try {
            in = new ObjectInputStream(new FileInputStream("data/objectFile.obj"));
            objectList = (ArrayList<GameObject>) in.readObject();
            in.close();
        } catch (FileNotFoundException e) {
            e.printStackTrace();
        } catch (IOException e) {
            e.printStackTrace();
        } catch (ClassNotFoundException e) {
            e.printStackTrace();
        }
        return objectList;
    }

    @Override
    public void initSprite() {
        final int SIZE = 8;
        screen = new Screen(SIZE);
        Random random = new Random();
        bullets = new HashSet<Plane.Bullet>();
        enemys = new HashSet<Enemy>();
        ArrayList<GameObject> objectList = readData();

        for(int i = 0; i < 10; i++) {
            Enemy enemy = new Enemy(random.nextInt(3),random.nextInt(7));
            enemys.add(enemy);
        }

        if(objectList == null) {
            plane = new Plane(7,4);
        }
        else {
            for(GameObject object:objectList) {
                if(object instanceof Plane) {
                    plane = (Plane) object;
```

```java
            }
            if(object instanceof Plane.Bullet bullet) {
                bullets.add(bullet);
            }
        }
    
    screen.add(bullets);
    screen.add(plane);
    screen.add(enemys);
}

@Override
public void update() {
    char key = screen.getKey();
    screen.delay();
    
    if(key == 'w') {
        plane.moveUp();
    }
    if(key == 's') {
        plane.moveDown();
    }
    if(key == 'a') {
        plane.moveLeft();
    }
    if(key == 'd') {
        plane.moveRight();
    }
    
    if(key == 'k') { // 发射子弹
        bullet = plane.shoot();
        screen.add(bullet);
        bullets.add(bullet);
    }
    
    if(key == 'p') { //保存游戏数据
        try {
            saveData(screen.getData());
        } catch (Exception e) {
            e.printStackTrace();
        }
    }
}
}
```

运行程序，屏幕上显示出大量的敌机。

接下来实现子弹击中敌机之后，敌机消失的功能。子弹可能击中屏幕上任何一个敌机，所以需要遍历保存敌机的 HashSet 集合，如果子弹与集合中某一个敌机相撞(位置发生重合)，则子弹和该敌机都消失，代码如下：

```java
public class Plane extends GameObject implements Move{
    public Plane(int r, int c) {
        super(r,c);

    }

    @Override
    public void moveUp() {
        if(getRow() > 0) {
            setRow(getRow() - 1);
        }
    }

    @Override
    public void moveDown() {
        if(getRow() + 1 < Config.SCREENSIZE) {
            setRow(getRow() + 1);
        }
    }

    @Override
    public void moveLeft() {
        if(getCol() > 0) {
            setCol(getCol() - 1);
        }
    }

    @Override
    public void moveRight() {
        if(getCol() + 1 < Config.SCREENSIZE) {
            setCol(getCol() + 1);
        }
    }

    public Bullet shoot(HashSet<Enemy> enemys) {
        Bullet bullet = new Bullet(getRow(),getCol());
        bullet.enemys = enemys;
        Thread thread = new Thread(bullet);
        thread.start();
        return bullet;
    }

    public class Bullet extends GameObject implements Runnable{
        private int speed = 1;
        private boolean state = true;
        private HashSet<Enemy> enemys;
        public Bullet(int r, int c) {
            super(r,c);
        }

        public void setSpeed(int speed) {
            this.speed = speed;
        }
```

```java
        public int getSpeed() {
            return speed;
        }

        public void moveUp() {
            setRow(getRow() - 1);
        }

        public void setState(boolean state) {
            this.state = state;
        }

        public boolean isHit(GameObject object) {      //判断子弹是否击中敌机
            if(object.getRow() == getRow() && object.getCol() == getCol()) {
                object.setLife(false);
                setLife(false);
                return true;
            }
            else {
                return false;
            }
        }

        public void hit(HashSet<Enemy> enemys) {
            Iterator<Enemy> it = enemys.iterator();
            while(it.hasNext()){                       //遍历敌机,如果被子弹击中,则敌机消亡
                Enemy enemy = it.next();
                if(isHit(enemy)){
                    it.remove();
                }
            }
        }

        public void run() {
            state = getLife();
            while(state) {
                moveUp();
                hit(enemys);
                if(getRow() < 0) {
                    setLife(false);
                }

                try {
                    Thread.sleep(500 / speed);
                }catch(Exception e) {
                }
                if(speed < 5) {
                    speed++;
                }
            }
        }
    }
}
```

PlaneGame 类也需要做细微调整，调用 shoot() 方法需要传递参数，代码如下：

```java
public class PlaneGameNew extends GameCore{
    private Screen screen;
    Plane plane;
    HashSet<Plane.Bullet> bullets = new HashSet<Plane.Bullet>();
    HashSet<Enemy> enemys = new HashSet<Enemy>();
    Plane.Bullet bullet;

    public void saveData(ArrayList<GameObject> objectList) {
        ObjectOutputStream out;
        try {
            out = new ObjectOutputStream(new FileOutputStream("data/objectFile.obj"));
            out.writeObject(objectList);
            out.close();
        } catch (FileNotFoundException e) {
            e.printStackTrace();
        } catch (IOException e) {
            e.printStackTrace();
        }
    }

    public ArrayList<GameObject> readData() {
        ArrayList<GameObject> objectList = null;
        ObjectInputStream in;
        try {
            in = new ObjectInputStream(new FileInputStream("data/objectFile.obj"));
            objectList = (ArrayList<GameObject>) in.readObject();
            in.close();
        } catch (FileNotFoundException e) {
            e.printStackTrace();
        } catch (IOException e) {
            e.printStackTrace();
        } catch (ClassNotFoundException e) {
            e.printStackTrace();
        }
        return objectList;//读取 customer 对象
    }

    @Override
    public void initSprite() {
        final int SIZE = 8;
        screen = new Screen(SIZE);
        Random random = new Random();
        bullets = new HashSet<Plane.Bullet>();
        enemys = new HashSet<Enemy>();
        ArrayList<GameObject> objectList = readData();

        for(int i = 0; i < 10; i++) {
            Enemy enemy = new Enemy(random.nextInt(3),random.nextInt(7));
            enemys.add(enemy);
        }
```

```java
            if(objectList == null) {
                plane = new Plane(7,4);
            }
            else {
                for(GameObject object:objectList) {
                    if(object instanceof Plane) {
                        plane = (Plane) object;
                    }
                    if(object instanceof Plane.Bullet bullet) {
                        bullets.add(bullet);
                    }
                }
            }

            screen.add(bullets);
            screen.add(plane);
            screen.add(enemys);

        }

        @Override
        public void update() {
            char key = screen.getKey();
            screen.delay();

            if(key == 'w') {
                plane.moveUp();
            }
            if(key == 's') {
                plane.moveDown();
            }
            if(key == 'a') {
                plane.moveLeft();
            }
            if(key == 'd') {
                plane.moveRight();
            }
            if(key == 'k') {
                bullet = plane.shoot(enemys);
                screen.add(bullet);
                bullets.add(bullet);
            }

            if(key == 'p') {
                try {
                    saveData(screen.getData());
                } catch (Exception e) {
                    e.printStackTrace();
                }
            }
        }
    }
```

测试代码如下：

```
public class Main {
    public static void main(String[ ] args){
        PlaneGame game = new PlaneGame();
        game.init();
        game.run();
    }
}
```

运行程序，按下按键"k"的时候，飞机发射子弹。当子弹击中敌机的时候，敌机会消失。

习题

11.1 当线程调用 start()方法后，其所处的状态为_____。
    A. 阻塞状态　　　　B. 运行状态　　　　C. 就绪状态　　　　D. 新建状态

11.2 下列关于线程优先级的说法中，不正确的是_____。
    A. 线程的优先级是不可以改变的
    B. 线程的优先级是在创建线程时设置的
    C. 在创建线程后的任何时候都可以重新设置
    D. 线程的优先级的范围为 1~10

11.3 下列关于 Thread 类的线程控制方法的说法中错误的是_____。
    A. 线程可以通过调用 sleep()方法使比当前线程优先级低的线程运行
    B. 线程可以通过调用 yield()方法使和当前线程优先级一样的线程运行
    C. 线程的 sleep()方法调用结束后，该线程进入运行状态
    D. 若没有相同优先级的线程处于可运行状态，线程调用 yield()方法时，当前线程将继续执行

11.4 编写程序，实现游戏角色的跳跃功能。

第12章

数据库编程

使用Java语言开发系统程序的过程中,通常会使用数据库存储数据。比如,大型网络游戏中的用户管理系统,各种数据都是采用数据库存储。Java程序主要通过JDBC(Java Database Connectivity,Java 数据库连接)访问数据库。JDBC 是 Java 程序访问数据库的应用程序接口,由一组类和接口组成,它们是连接数据库和 Java 程序的桥梁。通过 JDBC,Java 程序可以方便地对各种主流数据库进行操作。

12.1 JDBC 概述

实际项目开发中,广泛使用的数据库管理系统有 MySQL、Oracle、SQL Server 等。这些主流的数据库管理系统由不同厂商生产,每一家数据库厂商采用的内部数据处理方式不同,它们都有自己的数据库访问接口。设计程序的时候,如果直接使用数据库厂商提供的访问接口操作数据库,那么程序的可移植性就变得很差,从一种数据库管理系统切换到另一种数据库管理系统将会非常复杂。而 JDBC 制定了访问各类关系数据库的标准接口,为各个数据库厂商提供了标准接口的实现,在应用程序与数据库之间架起了一座桥梁。应用程序不必直接与数据库联系,而是通过 JDBC 访问数据库,由 JDBC 和具体的数据库驱动联系,从而增强代码的通用性。JDBC 的工作原理如图 12.1 所示。

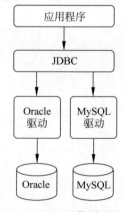

图 12.1 JDBC 工作原理示意图

12.2 JDBC 使用步骤

使用 JDBC 应用程序接口连接和访问数据库的方式较为固定，一般分为如下 5 个步骤：
(1) 加载驱动程序。
(2) 建立连接对象。
(3) 创建语句对象。
(4) 获取 SQL 语句执行结果。
(5) 关闭对象，释放资源。

12.2.1 加载驱动程序

访问数据库时，首先要加载数据库驱动，其目的是告诉程序将要连接哪个厂商的数据库。数据库驱动是不同数据库厂商为了实现统一的数据库调用而开发出的程序，对于 Java 语言来说，数据库驱动就是 java.sql.Driver 接口的实现类。由于驱动本质上还是一个类，加载数据库驱动和加载普通类原理一样，可以使用 Class 类的 forName() 静态方法，具体格式如下：

```
Class.forName(className);
```

该方法返回一个 Class 类的对象。其中，参数 className 就是数据库驱动类名称所对应的字符串。

例如，加载 Oracle 数据库驱动程序的代码如下：

```
Class.forName("oracle.JDBC.driver.OracleDriver");
```

加载 SQL Server 数据库驱动程序的代码如下：

```
Class.forName("com.microsoft.JDBC.sqlserver.SQLServerDriver");
```

加载 MySQL 数据库驱动程序的代码如下：

```
Class.forName("com.mysql.JDBC.Driver");
```

加载驱动程序时，需要确保已经成功安装了对应数据库的驱动程序。

12.2.2 建立连接对象

创建与数据库的连接是通过 DriverManager 类的 getConnection() 静态方法来实现。通过 DriverManager 类获取数据库连接的方式如下：

```
Connection connection = DriverManager.getConnection(String url, String user, String pwd);
```

getConnection() 方法的作用是获得 Connection 接口对象，只有获得该连接对象后，才能访问数据库，并操作数据表。getConnection() 方法里面的三个参数分别是连接数据库的 URL、登录数据库的用户名和密码。其中,用户名和密码通常由数据库管理员设置,而 URL 用于标识数据库的位置,URL 的书写遵循一定的规则,格式如下:

```
协议名:子协议名:数据源名
```

其中:
(1) JDBC URL 的协议名总是"jdbc"。
(2) 子协议名表示数据库类型协议,即驱动程序或者数据库连接机制的名称。
(3) 数据源名为数据库标识符,一般包括主机名、端口和数据库名等信息。

例如,MySQL 数据库的 URL 地址的写法示例如下:

```
jdbc:mysql://localhost:3306/Test
```

其中,jdbc:mysql 是固定写法,mysql 指的是 MySQL 数据库;localhost 表示数据库安装在本机,如果不在本机,则是所要连接机器的 IP 地址;数字 3306 是连接数据库的端口号,MySQL 端口号默认为 3306;Test 是数据库的名称。

Oracle 数据库的 URL 地址的写法示例如下:

```
jdbc:oracle:thin:@localhost:1521:Test
```

SqlServer 数据库的 URL 地址的写法示例如下:

```
jdbc:microsoft:sqlserver://localhost:1433;DatabaseName=Test
```

通过示例可知,对于选定的数据库,协议名和子协议名是固定的,唯一有变化的是数据源名。
成功创建连接后会返回 Connection 对象,通过该连接对象可以对数据库进行各种操作。

12.2.3 创建语句对象

得到 Connection 对象后,可以调用该对象的方法创建 SQL(Structured Query Language,结构化查询语言)语句对象,并通过其执行 SQL 语句。SQL 语句是对数据库进行操作的一种语言,用于存取、查询、更新和管理关系数据库中的数据。执行 SQL 语句的对象有 Statement、PreparedStatement 和 CallableStatement 三种类型。其中,Statement 提供了执行语句和获取结果的基本方法;PreparedStatement 添加了处理输入参数的方法;CallableStatement 支持调用存储过程,提供了对输出输入参数的支持。

Statement 对象每次执行 SQL 语句都需要编译 SQL 语句,通常适用于仅执行一次查询并返回结果的情形。PreparedStatement 对象执行 SQL 语句会预编译,只需要编译一次就能多次执行,能够有效提高系统性能,适用于多次使用 SQL 语句的场景。CallableStatement 对象主要调用数据库中的存储过程,存储过程是在大型数据库系统中,存储在数据库中的一组为了完成特定功能的 SQL 语句集。用户可以通过指定存储过程的名字来执行它,在数据量非常大的情况下,使用存储过程可以极大提升效率。

可以通过调用 Connection 接口对象的相应方法得到上述三种对象,具体方法如下:
(1) createStatement():创建基本的 Statement 对象。

（2）prepareStatement(String sql)：根据 SQL 语句创建 prepareStatement 对象。

（3）prepareCall (String sql)：根据 SQL 语句创建 CallableStatement 对象。

例如：

```
Statement stmt = connection.createStatement();
String sql = "update tb_user set password = '123456' where name = 'Tom'";
prepareStatement stmt = connection.prepareStatement(sql);
```

所有的 Statement 对象都有如下三种执行 SQL 语句的方法。

（1）execute(String sql)：用于执行任意 SQL 语句，该方法的返回值为 boolean 类型，作用是判断第一个结果是否为 ResultSet 对象，如果是则返回"true"，否则返回"false"。

（2）executeQuery(String sql)：用于执行查询语句，该方法的返回值为 ResultSet 结果集对象。

（3）executeUpdate (String sql)：主要用于执行 DML(数据操作语言)以及 DDL(数据定义语言)语句，该方法的返回值为一个整数，含义是受操作影响的记录数。DML 语句为 INSERT、UPDATE 或 DELETE，DDL 语句如 CREATE TABLE 或 DROP TABLE。

例如：

```
String sql = "select * from tb_user";
ResultSet rs = stmt.executeQuery(sql);
```

12.2.4　获取 SQL 语句执行结果

如果执行的 SQL 语句是查询语句，执行结果将返回一个 ResultSet 对象。ResultSet 对象表示 SQL 查询语句获得的结果集，该对象保存了 SQL 语句的查询结果，使用该对象可以将查询结果显示出来。例如，ResultSet 对象中的 next()方法用于定位结果集中的下一条记录，该方法返回值为"true"或者"false"。如果返回结果为"true"则表示成功定位一条记录，此时可通过 getString()等方法对该记录进行访问；如果返回结果为"false"，表示已无记录。

例如，从 tb_user 数据库表中获得所有用户的用户名和密码，并将其打印出来，代码如下：

```
String sql = "select * from tb_user";
ResultSet rs = stmt.executeQuery(sql);

while(rs.next()) {
    String name = rs.getString("name");
    String password = rs.getString("password");
    System.out.println(name + password);
}
```

12.2.5　关闭对象，释放资源

数据库访问结束后，应该关闭数据库连接，释放资源，以重复利用资源。

【例 12.1】 访问 MySQL 数据库，并且从表中读取数据，将结果打印在控制台上。完成任务之前，需要安装 MySQL 数据库管理系统和 MySQL 数据库驱动程序。

视频讲解

成功安装 MySQL 数据库管理系统和驱动程序之后,在 MySQL 数据库管理系统中创建数据库和数据库表。例如,创建一个名称为"usermanage"的数据库,然后在该数据库中创建一个名称为"tb_user"的表。创建"tb_user"表的时候,添加"name""password"字段。数据库和表创建成功后,再向表中插入 2 条数据记录。数据插入成功后,可以查看其中的数据。

简单查询 SQL 语句的基本语法格式如下:

```
SELECT * FROM 表名
```

完整代码如下:

```java
public class Main {
    public static void main(String[ ] args) throws Exception{
        Connection conn = null;
        Statement stmt = null;
        ResultSet rs = null;

        try{
            Class.forName("com.mysql.cj.jdbc.Driver");
            String url = "jdbc:mysql://localhost:3306/usermanage";
            conn = DriverManager.getConnection(url, "root", "root");
            stmt = conn.createStatement();
            String sql = "select * from tb_user";
            rs = stmt.executeQuery(sql);

            while(rs.next()) {
                String name = rs.getString("name");
                String password = rs.getString("password");
                System.out.println("name:" + name + "password:" + password);
            }
        }catch(Exception e) {
            e.printStackTrace();
        }finally {
            rs.close();
            stmt.close();
            conn.close();
        }
    }
}
```

程序运行结果如图 12.2 所示。

通过结果可知,成功访问数据库,并获得 tb_user 里的数据表信息。

【例 12.2】 访问 MySQL 数据库,并且向表中新增一条数据,然后查询数据库表中的数据,将结果打印在控制台上。

新增一条数据的 SQL 语句基本语法格式如下:

```
INSERT INTO 表名 (字段 1,字段 2,…) VALUES(值 1,值 2,…);
```

视频讲解

图 12.2　数据表显示示例

完整代码如下：

```java
public class Main {
    public static void main(String[ ] args) throws Exception{
        Connection conn = null;
        Statement stmt = null;
        ResultSet rs = null;
        
        try{
            Class.forName("com.mysql.cj.jdbc.Driver");
            String url = "jdbc:mysql://localhost:3306/usermanage";
            conn = DriverManager.getConnection(url, "root", "root");
            stmt = conn.createStatement();
            String sql = "insert into tb_user(name,password) values('Limei','123456')";
            stmt.executeUpdate(sql);
            
            String sqlQuery = "select * from tb_user";
            rs = stmt.executeQuery(sqlQuery);
            while(rs.next()) {
                String name = rs.getString("name");
                String password = rs.getString("password");
                System.out.println("name:" + name + " password:" + password);
            }
        }catch(Exception e) {
            e.printStackTrace();
        }finally {
            rs.close();
            stmt.close();
            conn.close();
        }
    }
}
```

程序运行结果如图 12.3 所示。

通过结果可知，成功新增一条数据到数据库中。

【例 12.3】 访问 MySQL 数据库，修改用户 Tom 的密码，并查询数据库表中的数据，将结果打印在控制台上。

视频讲解

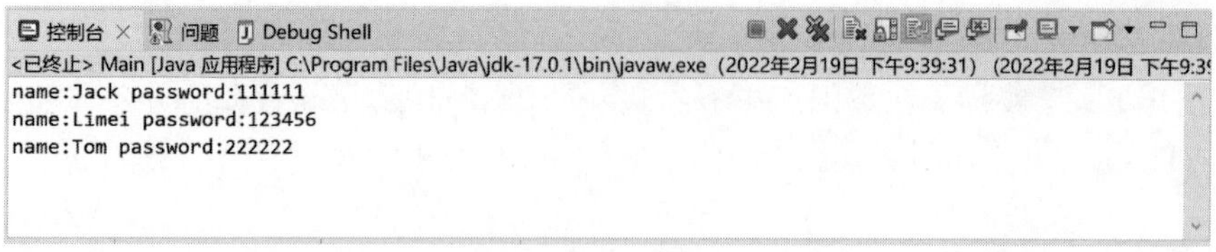

```
name:Jack password:111111
name:Limei password:123456
name:Tom password:222222
```

图 12.3　新增数据示例的运行结果

更新记录的 SQL 语句基本语法格式如下：

UPDATE 表名 SET 列名 = 表达式,列名 = 表达式 … WHERE 条件

代码如下：

```java
public class Main {
    public static void main(String[ ] args) throws Exception{
        Connection conn = null;
        PreparedStatement stmt = null;
        ResultSet rs = null;
        
        try{
            Class.forName("com.mysql.cj.jdbc.Driver");
            String url = "jdbc:mysql://localhost:3306/usermanage";
            conn = DriverManager.getConnection(url, "root", "root");
            
            String sql = "update tb_user set password = '123456' where name = 'Tom'";
            stmt = conn.prepareStatement(sql);
            stmt.executeUpdate();
            String sqlQuery = "select * from tb_user";
            rs = stmt.executeQuery(sqlQuery);
            while(rs.next()) {
                String name = rs.getString("name");
                String password = rs.getString("password");
                System.out.println("name:" + name + " password:" + password);
            }
        }catch(Exception e) {
            e.printStackTrace();
        }finally {
            rs.close();
            stmt.close();
            conn.close();
        }
    }
}
```

程序运行结果如图 12.4 所示。

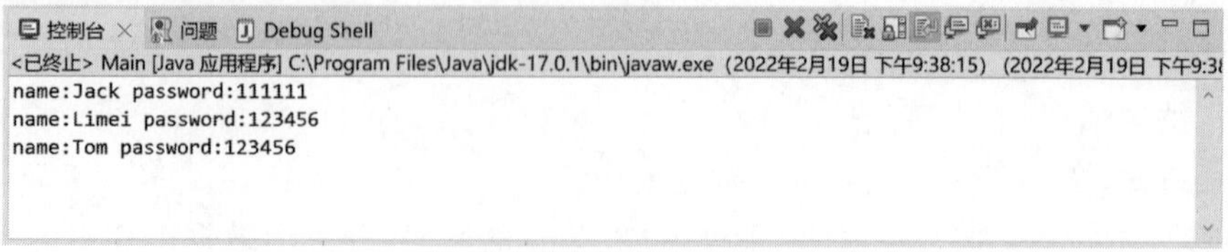

```
name:Jack password:111111
name:Limei password:123456
name:Tom password:123456
```

图 12.4　修改数据示例的运行结果

通过结果可知,用户 Tom 的密码成功由"222222"修改成"123456"。

例 12.1~例 12.3 演示了在数据库中进行简单查询、新增和更新数据的方法,观察代码会发现有大量重复的代码,可以设计一个数据库工具类 DBUtil,减少重复代码,代码如下:

```java
public class DBUtil {
    private static String driver = "com.mysql.cj.jdbc.Driver";
    private static String url = "jdbc:mysql://localhost:3306/usermanage";
    private static String user = "root";
    private static String password = "root";

    public static Connection getConnection() {                    //获取连接对象
        Connection conn = null;
        try {
            Class.forName(driver);
            conn = DriverManager.getConnection(url,user,password);
        }catch(Exception e) {
            e.printStackTrace();
        }
        return conn;
    }

    public static void close(ResultSet result, Statement stat, Connection conn) {//关闭对象,释放资源
        try {
            if(result != null) {
                result.close();
            }
            close(stat,conn);
        }catch(Exception e) {
            e.printStackTrace();
        }
    }

    public static void close(Statement stat, Connection conn) {
        try {

            if(stat != null) {
                stat.close();
            }

            if(conn != null) {
                conn.close();
            }
        }catch(Exception e) {
            e.printStackTrace();
        }
    }
}
```

测试代码如下:

```java
public class Main {
    public static void main(String[ ] args) throws Exception{
        Connection conn = DBUtil.getConnection();
```

```java
        PreparedStatement  stmt = null;
        ResultSet rs = null;

        try{
            String sql = "update tb_user set password = '000000' where name = 'Tom'";
            stmt = conn.prepareStatement(sql);
            stmt.executeUpdate();
            String sqlQuery = "select * from tb_user";
            rs = stmt.executeQuery(sqlQuery);
            while(rs.next()) {
                String name = rs.getString("name");
                String password = rs.getString("password");
                System.out.println("name:" + name + " password:" + password);
            }
        }catch(Exception e) {
            e.printStackTrace();
        }finally {
            DBUtil.close(rs,stmt,conn);
        }
    }
}
```

程序运行结果如图 12.5 所示。

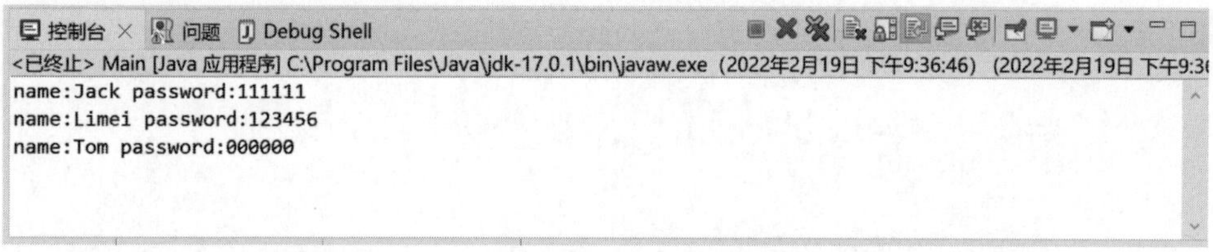

```
name:Jack password:111111
name:Limei password:123456
name:Tom password:000000
```

图 12.5　数据库工具类示例的运行结果

通过结果可知,用户 Tom 的密码成功由"123456"修改成"000000"。

利用 DBUtil 类操作数据库,可以避免每次重复进行加载驱动程序、建立连接、关闭数据库等程序的编写。

12.3　DAO 设计模式

在实际项目开发中,应该将所有对数据源的访问操作进行抽象化封装起来,提供公共应用程序接口(Application Program Interface,API)进行访问。数据访问对象(Data Access Object,DAO)模式就是基于此思想的一种设计模式,它提供了访问关系型数据库系统所需要的操作接口,将数据访问代码从业务逻辑代码中分离出来,实现分层设计,降低代码间的耦合度,提高代码的扩展性和系统可移植性。DAO 模式的优势在于将数据访问层的代码集中到一起,远离业务逻辑代码,这样在不影响应用中其他层的前提下能够独立开发、优化。另外,DAO 设计模式可以避免应用受到来自持久层的改变所带来的影响。例如,将项目的数据库管理系统从 Oracle 换到 MySQL,只需要做微小的调整就能实现。

一个典型 DAO 模式的组成包括 DAO 接口、DAO 实现类、实体类、数据库连接和关闭工具类。

DAO 设计模式的核心是"Java"接口,在里面可以定义各种 DAO 操作。如增加、修改、删除等方法。

下面使用 DAO 模式重新设计例 12.2 和例 12.3,读者可以直观感受 DAO 模式的作用。其中,数据库连接和关闭工具类 DBUtil 已经设计好了,接下来只需要设计实体类、DAO 接口和实现类即可完成该任务。

设计实体类 User,该类对象主要存储用户基本信息,代码如下:

```java
public class User {
    private String username;
    private String password;
    public String getUsername() {
        return username;
    }
    public void setUsername(String username) {
        this.username = username;
    }
    public String getPassword() {
        return password;
    }
    public void setPassword(String password) {
        this.password = password;
    }
    public User(String username, String password) {
        super();
        this.username = username;
        this.password = password;
    }
    public User() {
        super();
    }
}
```

Dao 接口 UserDao,定义了两个抽象方法,向数据库中增加和更新用户信息,代码如下:

```java
public interface UserDao {
    public void addUser(User user) ;
    public void updateUser(User user);
}
```

Dao 实现类 UserDaoImpl,具体实现了数据库操作的常用方法,代码如下:

```java
public class UserDaoImpl implements UserDao{
    public void addUser(User user) {
        Connection conn = null;
        PreparedStatement pst = null;
        conn = DBUtil.getConnection();
        String sql = "insert into tb_user(name, password) values(?,?)";

        try {
```

```java
                pst = conn.prepareStatement(sql);
                pst.setString(1, user.getUsername());
                pst.setString(2, user.getPassword());
                pst.executeUpdate();
            } catch (SQLException e) {
                e.printStackTrace();
            }
        }

        public void updateUser (User user) {
            Connection conn = null;
            PreparedStatement pst = null;
            conn = DBUtil.getConnection();
            String sql = "update tb_user set password = ? , name = ?";
            try {
                pst = conn.prepareStatement(sql);
                pst.setString(1, user.getUsername());
                pst.setString(2, user.getPassword());
                pst.executeUpdate();
            } catch (SQLException e) {
                e.printStackTrace();
            }
        }
    }
```

测试程序的代码如下:

```java
public class Main {
    public static void main(String[] args) throws Exception{
        UserDaoImpl dao = new UserDaoImpl();
        User user = new User("Lily", "123456");
        dao.addUser(user);
    }
}
```

运行程序,查看数据库可知用户新增成功。

通过这个例子可知,使用 DAO 模式将数据访问分为抽象层和实现层,将数据使用和数据访问相分离,数据访问层代码发生变化也不会影响到业务逻辑代码。例如,上述例子中将数据库管理系统从 MySQL 系统切换到 Oracle 系统,只需要修改 DBUtil 类即可,无须修改其他代码。

12.4 综合案例：用户管理系统

用户管理是每一个后台产品必备模块,比如每一款大型游戏都需要一个用户管理系统。用户管理系统可以让用户更好地了解游戏、享受游戏,也方便游戏管理员管理用户,提高用户体验。

编写程序,实现用户管理系统最基本的账户管理功能,包括注册、登录功能。注册功能用于注册新用户。登录功能保证用户只有输入正确的用户名和密码时才能登录成功,从而进入游戏界面开始玩游戏。

实现注册和登录功能的关键是访问和操作数据库。注册过程是将新注册的用户信息到数据库中查询,检测该用户名是否已经被注册过。如果已经被注册过,则返回失败信息;如果没有被注册,则将用户名和密码添加到数据库中,并返回注册成功信息。登录过程是根据登录的用户名信息去数据库中查询,将查询结果集与密码进行比对,如果一致,则登录成功,否则登录失败。

程序的设计采用 DAO 模式,对应的实体类、数据库连接和关闭工具类、DAO 接口和实现类代码分别如下。

(1) 用户类,主要成员变量包括用户名、密码,主要的成员方法有获得用户名、获得密码、设置用户名和设置密码,代码如下:

```java
public class User {
    private String username;
    private String password;
    public String getUsername() {
        return username;
    }
    public void setUsername(String username) {
        this.username = username;
    }
    public String getPassword() {
        return password;
    }
    public void setPassword(String password) {
        this.password = password;
    }
    public User(String username, String password) {
        super();
        this.username = username;
        this.password = password;
    }
    public User() {
        super();
    }
}
```

(2) 数据库连接和工具类,代码如下:

```java
public class DBUtil {
    private static String driver = "com.mysql.cj.jdbc.Driver";
    private static String url = "jdbc:mysql://localhost:3306/usermanage";
    private static String user = "root";
    private static String password = "root";


    public static Connection getConnection() {                    //获取连接对象
        Connection conn = null;
        try {
            Class.forName(driver);
            conn = DriverManager.getConnection(url,user,password);
        }catch(Exception e) {
```

```java
            e.printStackTrace();
        }
        return conn;
    }
    public static void close(ResultSet result, Statement stat, Connection conn) {//关闭对象,释放资源
        try {
            if(result != null) {
                result.close();
            }
            close(stat,conn);
        }catch(Exception e) {
            e.printStackTrace();
        }
    }

    public static void close(Statement stat, Connection conn) {
        try {

            if(stat != null) {
                stat.close();
            }

            if(conn != null) {
                conn.close();
            }
        }catch(Exception e) {
            e.printStackTrace();
        }
    }
}
```

(3) UserDao 接口,主要定义注册和登录功能的抽象方法,代码如下:

```java
public interface UserDao {
    public boolean regist(User user) ;                              //注册功能
    public boolean isLogin(String userName,String passWord);        //登录功能
}
```

(4) 用户操作类,也就是 UserDao 接口的实现类,代码如下:

```java
public class UserDaoImpl implements UserDao{
    public boolean regist(User user) {
        boolean result = false;
        Connection conn = null;
        PreparedStatement pst = null;
        conn = DBUtil.getConnection();
        String sql = "insert into tb_user(name, password) values(?,?)";

        try {
            pst = conn.prepareStatement(sql);
            pst.setString(1, user.getUsername());
            pst.setString(2, user.getPassword());
            int row = pst.executeUpdate();
```

```java
            if(row > 0){
                result = true;
            }
        } catch (SQLException e) {
            e.printStackTrace();
        } finally {
            DBUtil.close(pst, conn);
        }
        return result;
    }

    public boolean isLogin(String userName,String passWord) {
        boolean flag = false;
        Connection conn = null;
        Statement st = null;
        ResultSet rs = null;

        //创建 Connnection, Statement, ResultSet 对象并调用验证登录的方法
        String sql = "select * from tb_user where name = '" + userName + "'";
        conn = DBUtil.getConnection();
        try {
            st = conn.createStatement();
            rs = st.executeQuery(sql);
            while (rs.next()){
                if(rs.getString("password").equals(passWord)){
                    flag = true;
                }
            }
        } catch (SQLException e) {
            e.printStackTrace();
        } finally {
            DBUtil.close(rs, st, conn);
        }
        return flag;
    }
}
```

测试类,测试注册和登录功能,代码如下:

```java
public class Main {
    public static void main(String[ ] args) throws Exception{
        String userName = JOptionPane.showInputDialog("请输入用户名:");
        String password = JOptionPane.showInputDialog("请输入密码:");

        UserDaoImpl ud = new UserDaoImpl();
        User user = new User();
        user.setUsername(userName);
        user.setPassword(password);
        boolean flag = ud.regist(user);
        if(flag) {
            JOptionPane.showMessageDialog(null,"注册成功");
        }
```

```java
        else{
            JOptionPane.showMessageDialog(null,"注册失败");
        }

        userName = JOptionPane.showInputDialog("请输入用户名:");
        password = JOptionPane.showInputDialog("请输入密码:");

        boolean result = ud.isLogin(userName,password);

        if(result){
            JOptionPane.showMessageDialog(null,"登录成功");
        }else{
            JOptionPane.showMessageDialog(null,"登录失败");
        }
    }
}
```

运行程序,测试注册、登录是否成功。

本章通过案例让读者掌握了使用 JDBC 开发数据库应用程序的基础知识。数据库是信息化核心环节,大数据时代的来临、海量数据的涌现,使数据库领域高速发展,如传统集中式数据库向分布式架构升级、非关系数据库快速发展等,国家新基建对数据库方面的人才需求也在不断扩大。本章内容只介绍了 Java 程序访问数据库的基础知识,起到抛砖引玉的作用,读者如果对该领域感兴趣可以深入研究。

习题

12.1 下面描述中错误的是_____。

A. Statement 对象的 executeQuery()方法会返回一个结果集。

B. Statement 对象的 execute()方法会返回一个 boolean 类型的值,含义是判断第一个结果是否为 ResultSet 对象。

C. Statement 对象的 executeUpdate()方法会返回一个 boolean 类型的值,含义是判断更新是否成功。

D. Statement 对象的 executeUpdate()方法会返回一个 int 类型的值,含义是计算受操作影响的记录数。

12.2 以下负责建立与数据库连接的对象是_____。

A. Statement   B. PreparedStatement

C. ResultSet   D. DriverManager

12.3 使用 Connection 对象的_____方法可以建立一个 PrepareStatement 接口对象。

A. createPrepareStatement()   B. prepareStatement()

C. prepareCall()   D. createStatement()

12.4 编写程序,实现 QQ 账号管理系统。

第13章

网络编程

如今的生活已经离不开网络,移动支付、无人驾驶、物联网等技术彻底改变了人们的生活。Java 语言的平台无关性和强大的网络支持功能,使得 Java 语言在互联网时代大放异彩,成为最热门的语言之一。网络编程的主要目的就是实现计算机之间的通信,使数据在不同设备之间传输。

13.1 网络通信概述

计算机之间的通信是一个复杂的过程,涉及传输介质、通信设施和网络通信协议等。网络程序设计就是编写程序使数据在不同的设备之间进行传输。由于 Java 语言将网络程序所需要的内容封装成不同的类,使得程序设计者即使不熟悉网络相关的知识,也能通过调用标准库提供的接口高效地编写网络程序。

编写网络程序,首先要确定应用程序所使用的网络通信协议。网络通信协议就是计算机在通信过程中必须遵守的规则。通信协议对速率、传输代码、代码结构、传输控制步骤、出错控制等制定了标准。进行网络通信时,通信双方必须遵守通信协议,只有这样双方才能进行数据交换,这就像商品交易时必须遵守商业规则一样。网络通信协议有许多种,应用最广泛的是 TCP/IP 网络通信协议。

TCP/IP 是一个协议族,包括传输控制协议(Transmission Control Protocol,TCP)、网络互联协议(Internet Protocol,IP)和用户报文协议(User Datagram Protocol,UDP)等多个具有不同功能且互为关联的协议。java.net 包中提供了 TCP 和 UDP 两种常见的网络协议的支持。TCP 协议是一种面向连接的传输层控制协议,即在传输数据前先在发送端和接收端建立逻辑连接,然后再传输数据,它提供了两台计算机之间可靠无差错的数据传输。UDP 是用户报文协议,是无连接通信协议,即在数据传输时,发送端和接收端不需要建立逻辑连接。当一台计算机向另外一台计算机发送数据时,发送端发出数据时不会确认接收端是否存在。同样,接收端在收到数据时,也不会向发送端反馈是否收到数据,简而言之就是 UDP 只负责发送,不保证能否收到。两者都是传输层的协议,主要区别在于前者是可靠的,面向连接的协议,后者是不可靠的,无连接的协议。

TCP 遵循客户端与服务器端模式,也就是经典的 C/S 模式。在通信时,客户端去连接服务器端,建立连接之后实现通信,服务器端不可以主动连接客户端。UDP 不需要建立连接,所以不区分客户端和服务器端,只分发送端和接收端,可以任意地发送数据。

针对网络通信的不同层次,Java 提供的网络功能有如下几类:

(1) InetAddress:用于标识网络上的硬件资源。

(2) Socket:使用 TCP 实现网络通信的 Socket 相关的类。

(3) Datagram:使用 UDP,将数据保留在数据报中,通过网络进行通信。

【例 13.1】 编写程序获得本机的计算机名和 IP 地址。

InetAddress 类中的获得计算机名和 IP 地址的方法,分别为 getHostName()方法和 getHostAddress()方法,代码如下:

```java
public class Main {
    public static void main(String[ ] args) throws Exception{
        //获取本机的 InetAddress 实例
        InetAddress localHost = InetAddress.getLocalHost();
        System.out.println("计算机名:" + localHost.getHostName());
        System.out.println("IP 地址:" + localHost.getHostAddress());
    }
}
```

程序运行结果如图 13.1 所示。

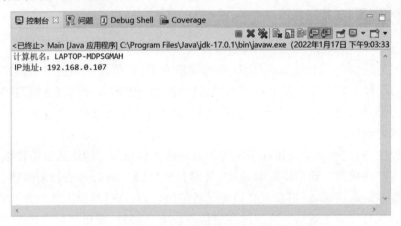

图 13.1　获取本机计算机名和 IP 地址结果图

13.2　TCP 通信

TCP 是一种可靠的网络协议,它在通信两端建立一个套接字(Socket)对象,从而在通信两端形成网络虚拟链路,两端的程序通过虚拟链路进行通信。Socket 是通信的基础技术,可以看作通信两端的收发器,整个过程与读写文件非常相似:读写文件的时候,首先要定位并打开文件,然后才能对数据进行读写。Socket 通信也类似,首先要通信双方建立连接,然后才能对数据进行收发。建立网络通信至少需要一对 Socket 对象,TCP 区分客户端和服务器端,Java 语言为客户端提供 Socket 类创建套接字对象,为服务器端提供了 ServerSocket 类创建套接字对象。

基于 TCP 的客户端和服务器端通信过程,首先是在服务器端创建一个服务器端套接字,并将它绑定在一个端口上,接着从这个端口监听连接。客户端请求与服务器端进行连接的时候,根据服务器的域名或者 IP 地址,加上端口号,打开一个套接字。其中,IP 地址是用来标识网络中的一个通信实体的地址,端口用来区分不同应用程序,每个应用程序都会有一个端口号,端口的范围为 0~65535,其中 0~1023 为系统保留,QQ、360 卫士等网络程序都有自己的端口。当服务器端接收到连接后,服务器端和客户端之间就能通过输入输出流进行数据交换。

【例 13.2】 模拟聊天程序,实现客户端与服务器端一对一聊天。

视频讲解

服务器端程序主要通过 java.net 包中 ServerSocket 类来标识服务器端。该类的 accept()方法用于监听是否有客户端发出请求,等待客户端的连接。如果没有客户端成功连接就一直处于阻塞状态,如果与客户端成功连接就返回一个与之对应的 Socket 对象,然后用该 Socket 对象与客户端通信。代码如下:

```java
public class Server {
    private int port = 8080;                  // 默认服务器端口
    public Server() {
    }

    public Server(int port) {                 // 创建指定端口的服务器
        this.port = port;
    }

    public void service() {                   // 提供服务
        try {                                 // 建立服务器连接
            ServerSocket server = new ServerSocket(port);
            System.out.println("服务器:启动,等待客户端的连接***");
            Socket socket = server.accept();  // 等待客户连接
            System.out.println("服务器:客户端已经连接成功!");
            try {
                DataInputStream in = new DataInputStream(socket.getInputStream());
                DataOutputStream out = new
                DataOutputStream(socket.getOutputStream());

                while (true) {
                    // 读取来自客户端的信息
                    String accpet = in.readUTF();
                    System.out.println("收到客户端信息:" + accpet);
                    //把服务器端的输入信息发给客户端
                    out.writeUTF("服务器端收到信息");
                }
            } finally {                       // 建立连接失败的话不会执行 socket.close();
                socket.close();
            }
        } catch (IOException e) {
            e.printStackTrace();
        }
    }
}
```

```java
        public static void main(String[ ] args) {
            new Server().service();
        }
    }
```

创建客户端程序,用于向指定服务器端发送连接并进行数据交互,客户端代码如下:

```java
public class Client {
    private String host = "localhost";              // 默认连接到本机
    private int port = 8080;                        // 默认连接到端口 8080
    public Client() {
    }

    public Client(String host, int port) {          // 连接到指定的主机和端口
        this.host = host;
        this.port = port;
    }

    public void chat() {
        try {
            Socket socket = new Socket(host, port); // 连接到服务器
            try {
                // 读取服务器端传过来信息的 DataInputStream
                DataInputStream in = new DataInputStream(socket
                        .getInputStream());
                // 向服务器端发送信息的 DataOutputStream
                DataOutputStream out = new DataOutputStream(socket
                        .getOutputStream());

                // 装饰标准输入流,用于从控制台输入
                BufferedReader br = new BufferedReader(new InputStreamReader(System.in));

                while (true) {
                    String send = br.readLine();
                    System.out.println("客户端:" + send);
                    out.writeUTF(send);              // 把从控制台得到的信息传送给服务器
                    String accpet = in.readUTF();
                    System.out.println(accpet);
                }
            } finally {
                socket.close();
            }
        } catch (IOException e) {
            e.printStackTrace();
        }
    }

    public static void main(String[ ] args) {
        new Client().chat();
    }
}
```

运行程序,客户端与服务器端之间能相互通信。

【例 13.3】 模拟聊天程序,实现多个客户端与服务器端通信。

例 13.2 中实现了客户端与服务器端的通信程序,但是只能实现一对一的通信。实际开发中,服务器端程序可以被多个客户端程序访问。要实现多个用户同时访问,服务器端可采用多线程技术实现。

视频讲解

服务器端线程类代码如下:

```java
public class ServerThread implements Runnable{
    private DataInputStream dis = null;
    private DataOutputStream dos = null;
    private Socket socket = null;

    public ServerThread(Socket socket) {
        this.socket = socket;
    }

    public void run() {
        try {
            while(true) {
                dis = new DataInputStream(socket.getInputStream());
                dos = new DataOutputStream(socket.getOutputStream());
                String message = dis.readUTF();
                System.out.println("服务器收到消息:" + message);
            }
        } catch (IOException e) {
            e.printStackTrace();
        } finally {
            try {
                socket.close();
            } catch (IOException e) {
                e.printStackTrace();
            }
        }
    }
}
```

服务器端类代码如下:

```java
public class Server {
    private int port = 8080;                    // 默认服务器端口
    public Server() {
    }

    public Server(int port) {                   // 创建指定端口的服务器
        this.port = port;
    }

    public void service(){
        try {
            ServerSocket server = new ServerSocket(port);
            System.out.println("服务器:启动,等待客户端的连接***");
```

```
        while(true) {                    // 等待客户连接
            Socket socket = server.accept();
            System.out.println("服务器:客户端已经连接成功!");
            ServerThread serverThread = new ServerThread(socket);
            new Thread(serverThread).start();
        }
    }catch (IOException e) {
        e.printStackTrace();
    }
}

public static void main(String[ ] args) {
    new Server().service();
}
}
```

客户端代码保持不变,运行服务器端程序之后,连续运行多次客户端程序模拟多客户端与服务器端建立连接并发送数据,发现多个客户端都可以与服务器端通信。服务器端与每一个客户端建立了连接后,就会为该客户端开一个新的线程进行数据通信。在实际项目中,这种方法只适用于少量客户端连接的情况。如果有大量客户端连接,需要使用线程池技术,以保证线程的复用。

13.3 UDP 通信

UDP 提供的服务不同于 TCP 的端到端服务,它是面向非连接的,属于不可靠协议,UDP 套接字在使用前不需要进行连接。UDP 不区分客户端和服务器端,只区分发送端和接收端,计算机之间可以任意地发送数据。Java 语言实现 UDP 通信主要通过 DatagramPacket 类和 DatagramSocket 类。DatagramPacket 类用来包装需要发送或者接收的数据,对其进行封装,就像运输快递时,要先将快递进行打包。DatagramSocket 类用来创建发送和接收数据包的套接字对象,发送信息时,Java 程序创建一个包含待发送信息的 DatagramPacket 对象,并将其作为参数传递给 DatagramSocket 对象的 send()方法。接收信息时,将 DatagramPacket 对象作为参数传递给 DatagramSocket 对象的 receive()方法。在创建 DatagramPacket 对象时用于发送端与接收端有所不同,如果该对象用来包装待接收的数据,则不指定数据来源的远程主机和端口,只需指定一个缓存数据的 byte 数组即可。而如果该对象用来包装待发送的数据,则要指定要发送到的目的主机和端口。这与生活中收发快递相似,寄快递一定要写明收信人地址信息,而收快递则不需要写明寄信人地址信息。

【例 13.4】 模拟聊天程序,实现接收端与发送端采用 UDP 发送信息。
采用 UDP 发送消息,接收端代码如下:

视频讲解

```
public class Receiver {
    private int port = 8800;                      // 默认服务器端口
    public Receiver() {
    }

    public Receiver(int port) {                   // 创建指定端口的服务器
```

```java
        this.port = port;
    }

    public void service(){                          // 提供服务
        DatagramSocket socket;
        try {
            socket = new DatagramSocket(port);
            byte[] data = new byte[1024];
            DatagramPacket packet = new DatagramPacket(data, data.length);
            while (true) {
                System.out.println("等待接收信息:");
                socket.receive(packet);              //在接收到数据报之前会一直阻塞
                String info = new String(data, 0, packet.getLength());
                System.out.println("收到信息:" + info);
            }
        } catch (IOException e1) {
            e1.printStackTrace();
        }
    }

    public static void main(String[] args) {
        new Receiver().service();
    }
}
```

发送端代码如下:

```java
public class Sender {
    public Sender() {
    }

    public void service(){ //提供服务
        try {
            InetAddress address = InetAddress.getByName("localhost");
            int port = 8800;
            // 装饰标准输入流,用于从控制台输入
            BufferedReader br = new BufferedReader(new InputStreamReader(System.in));
            byte[] data;

            // 创建 DatagramSocket 对象
            DatagramSocket socket = new DatagramSocket();
            // 向服务器端发送数据报
            while(true) {
                System.out.println("请输入信息:");
                data = br.readLine().getBytes();
                DatagramPacket packet = new DatagramPacket(data, data.length, address, port);
                socket.send(packet);
            }
        } catch (IOException e1) {
            e1.printStackTrace();
        }
    }
}
```

```java
        public static void main(String[ ] args) {
            new Sender().service();
        }
}
```

运行程序,可以实现发送端给接收端发送信息。

13.4 综合案例:网络版用户管理系统

网络版用户管理系统,相比于单机版用户管理系统,最大的区别就是信息需要通过网络传播,所以只需要新增网络通信相关的程序即可,访问数据库、注册和登录功能等程序无须修改。在本例中采用TCP,程序设计分为服务器端和客户端。客户端与服务器端通信程序的主要目标是进行数据交换,建立通信连接。进行数据交换的程序的基本流程都相似,不同的地方在于对接收信息的处理,所以可以设计一个通信抽象类,将处理信息的方法设计为抽象方法,代码如下:

```java
public abstract class ComThread implements Runnable{
    private DataInputStream dis = null;
    private DataOutputStream dos = null;
    private Socket socket = null;
    
    public ComThread(Socket socket) {
        this.socket = socket;
        try {
            dis = new DataInputStream(socket.getInputStream());
            dos = new DataOutputStream(socket.getOutputStream());
        } catch (IOException e) {
            e.printStackTrace();
        }
    }
    
    public void sendMessage(String message) {
        try {
            dos.writeUTF(message);
            dos.flush();
        } catch (IOException e) {
            e.printStackTrace();
        }
    }
    
    public void run() {
        while(true) {
            try {
                String message = dis.readUTF();
                System.out.println("收到服务器消息:" + message);
                dealwithMessage(message);
            }catch(IOException e) {
                e.printStackTrace();
```

```
            }
        }
    }

    public abstract void dealwithMessage(String message);
}
```

设计好通信抽象类后,客户端和服务器端的类都可以继承于该抽象类,实现抽象方法之后,建立连接进行数据交换。接下来就是分别实现这两个类的抽象方法 dealwithMessage(),这需要客户端与服务器端约定好数据交换格式,常用的数据格式有 Json、xml 和 html 类型。在本例中为了简便处理,自定义数据交换格式如下:

用户名:密码:行为

例如:

Tom:123456:login

表示用户名为 Tom,密码为 123456,执行登录行为。

根据数据交换的格式,则服务器端通信线程类的代码如下:

```
public class ServerThread extends ComThread{
    private UserDaoImpl userImp;
    public ServerThread(Socket socket, UserDaoImpl userImp) {
        super(socket);
        this.userImp = userImp;
    }

    @Override
    public void dealwithMessage(String message) {
        if(message.contains("register")) {                    //注册
            String name = message.split(":")[0];              //将信息拆分获得用户名
            String password = message.split(":")[1];          //将信息拆分获得密码
            User user = new User(name,password);

            if(userImp.regist(user)) {
                String res = "注册成功";
                sendMessage(res);
                System.out.println("注册成功");
            }
            else {
                String res = "注册失败";
                sendMessage(res);
                System.out.println("注册失败");
            }
        }

        if(message.contains("login")) {                       //登录
            String [] string = message.split(":");
            String name = string[0];
            String password = string[1];
```

```java
            System.out.println(name + password);
            if(userImp.isLogin(name,password)) {
                String res = "登录成功";
                sendMessage(res);
                System.out.println("登录成功");
            }
            else {
                String res = "登录失败";
                sendMessage(res);
                System.out.println("登录失败");
            }
        }
    }
}
```

其中,数据库相关的类 UserDaoImpl 类与第 12 章综合案例中的代码一致。

服务器端测试代码如下:

```java
public class Server {
    private int port = 8080;                          // 默认服务器端口
    private UserDaoImpl userImp;
    public Server() {
    }
    
    public Server(int port,UserDaoImpl userImp) {     // 创建指定端口的服务器
        this.port = port;
        this.userImp = userImp;
    }
    
    public void service(){
        try {
            ServerSocket server = new ServerSocket(port);
            System.out.println("服务器:启动,等待客户端的连接***");
            while(true) {                             // 等待客户连接
                Socket socket = server.accept();
                System.out.println("服务器:客户端已经连接成功!");
                ServerThread serverThread = new ServerThread(socket,userImp);
                new Thread(serverThread).start();
            }
        }catch (IOException e) {
            e.printStackTrace();
        }
    }
    
    public static void main(String[] args) {
        UserDaoImpl userDaoImpl = new UserDaoImpl();
        new Server(8080,userDaoImpl).service();
    }
}
```

客户端通信线程类代码如下:

```java
public class ClientThread extends ComThread{
    public ClientThread(Socket socket) {
        super(socket);
    }

    @Override
    public void dealwithMessage(String message) {
        if(message.contains("登录成功")) {
            JOptionPane.showMessageDialog(null,"登录成功");
        }
        if(message.contains("登录失败")) {
            JOptionPane.showMessageDialog(null,"登录失败");
        }
        if(message.contains("注册失败")) {
            JOptionPane.showMessageDialog(null,"注册失败");
        }
        if(message.contains("注册成功")) {
            JOptionPane.showMessageDialog(null,"注册成功");
        }
    }
}
```

客户端测试代码如下:

```java
public class Client {
    private String host = "localhost";              // 默认连接到本机
    private int port = 8080;                        // 默认连接到端口 8080
    private ClientThread thread;
    public Client() {
    }

    public Client(String host, int port) {          // 连接到指定的主机和端口
        this.host = host;
        this.port = port;
    }

    public void service(){
        try {
            Socket socket = new Socket(host,port);
            thread = new ClientThread(socket);
        }catch (IOException e) {
            e.printStackTrace();
        }
        new Thread(thread).start();
    }

    public void sendMessage(String message) {
        thread.sendMessage(message);
    }

    public static void main(String[ ] args) {
        Client client = new Client();
```

```
            client.service();

            String userName = JOptionPane.showInputDialog("请输入用户名:");
            String password = JOptionPane.showInputDialog("请输入密码:");

            String send = userName + ":" + password + ":register";
            client.sendMessage(send);

            userName = JOptionPane.showInputDialog("请输入用户名:");
            password = JOptionPane.showInputDialog("请输入密码:");
            send = userName + ":" + password + ":login";
            client.sendMessage(send);
        }
    }
```

分别运行服务器端、客户端程序,可以通过客户端进行注册或者登录。

网络程序设计还要考虑多线程、输入输出、数据报文粘包和分包等问题,一般会选择使用如 Netty、MINA 等已经非常成熟和稳定的网络编程框架。使用这些框架,只需要编写少量的代码,就能够实现高吞吐、低延时、高性能的高并发网络应用程序。读者如果感兴趣,可以对这些主流的开源框架进行深入学习。

习题

13.1 在 Java 网络编程中,使用客户端套接字 Socket 创建对象时,需要指定_____。
  A. 服务器地址和端口　　　　　　　　B. 服务器端口和文件
  C. 服务器名称和文件　　　　　　　　D. 服务器地址和文件

13.2 在使用 UDP 套接字通信时,常用_____类把要发送的信息打包。
  A. String　　　　　　　　　　　　　B. DatagramSocket
  C. MulticastSocket　　　　　　　　　D. DatagramPacket

13.3 Java 程序中,使用 TCP 套接字编写服务器端程序的套接字类是_____。
  A. Socket　　　　　　　　　　　　　B. ServerSocket
  C. DatagramPacket　　　　　　　　　D. DatagramSocket

13.4 编写程序,将一张图片从客户端传送到服务器端。

第三部分

综合应用篇

第14章

综合应用：网络版"飞机大战"

在学习了Java多线程、数据库和网络编程等知识后，可以尝试完成较为复杂的系统。

单机版"飞机大战"的主要模式是人机对战，而网络版"飞机大战"可以实现双打模式，邀请好友一起合作，共同面对同样的敌机、子弹等。双打模式需要双方紧密配合，相互合作，一起面对挑战，如图14.1所示。

网络双打模式中非常重要的一环就是相互之间的通信。比如，其中一方对飞机进行了操作，需要将这些消息传送到服务器上，并由服务器转发给另一个客户端，实现两个客户端之间数据同步。

通过服务器转发消息实现两个客户端通信的关键是：设置一个Hashmap对象保存每一个客户端的用户与

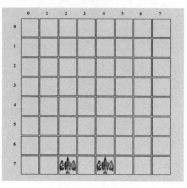

图14.1 网络版"飞机大战"游戏示意图

服务器端通信的线程之间的关系，这样就能通过服务器进行转发。另外，还需要设计好通信消息格式。例如，客户端向服务器端发送消息的格式约定如下：

发送方用户名:接收方用户名:数据:标识

例如：

LiLei:WangFang:moveUp:Action

意思是用户LiLei向WangFang发送了一个行为moveUp，Action的作用是一个标识，表示该消息是一个动作行为消息。

服务器端接收到信息后，解析出接收方用户名，通过Hashmap对象获得与该客户端对应的通信线程，使用该通信线程转发信息。为了实现客户端之间的通信，需要通信双方相互知道对方的用户名。

为了获得对方的用户名,用户登录的时候,需要向服务器发送登录信息,登录成功之后,服务器将群发新登录用户的信息,收到新登录的用户信息,通信双方建立联系。

服务器端 Server 类的代码如下:

```java
public class Server {
    HashMap<String,ServerThread> hashmapThread;
    private int port = 8080;                          // 默认服务器端口
    public Server() {
        hashmapThread = new HashMap<String,ServerThread>();
    }
    
    public Server(int port) {                         // 创建指定端口的服务器
        this.port = port;
        hashmapThread = new HashMap<String,ServerThread>();
    }
    
    public void service(){
        try {
            ServerSocket server = new ServerSocket(port);
            Socket socket;
            System.out.println("服务器:启动,等待客户端的连接***");
            while(true) {                             // 等待客户连接
                socket = server.accept();
                ServerThread serverThread = new ServerThread(socket);
                new Thread(serverThread).start();
            }
        }catch (IOException e) {
            e.printStackTrace();
        }
    }
}

public class ServerThread extends ComThread{
    public ServerThread(Socket socket) {
        super(socket);
    }
    
    @Override
    public void dealwithMessage(String message) {
        if(message.contains("connect")) {
            String name = message.split(":")[0];
            hashmapThread.put(name, this);
            
            for(String key:hashmapThread.keySet()) {
                ServerThread thread = hashmapThread.get(key);
                if(thread != this) {
                    thread.sendMessage(message);
                    this.sendMessage(key + ":connect");
                }
            }
```

```
            }
        }
    }
    public static void main(String[ ] args) {
        new Server(8080).service();
    }
}
```

服务器端的功能主要是接收数据和处理数据。当新用户登录的时候，服务器根据登录信息获得该用户的用户名，然后转发给已登录的客户端，并且将已登录的客户端的用户名转发给新登录的用户，这样双方可以建立通信。上述过程如同生活中使用微信通信一般，先互相加为好友，再进行通信。

客户端的代码如下：

```
public class Client {
    String name;
    private String host = "localhost";            // 默认连接到本机
    private int port = 8080;                      // 默认端口
    private ClientThread thread;
    public Client() {
        name = JOptionPane.showInputDialog("请输入用户名:");
    }

    public Client(String host, int port) {        // 连接到指定的主机和端口
        this.host = host;
        this.port = port;
    }

    public void service(){
        try {
            Socket socket = new Socket(host,port);
            thread = new ClientThread(socket);
            String send = name + ":connect";
            sendMessage(send);
        }catch (IOException e) {
            e.printStackTrace();
        }
        new Thread(thread).start();
    }

    public void sendMessage(String message) {
        thread.sendMessage(message);
    }

    public class ClientThread extends ComThread{
        public ClientThread(Socket socket) {
            super(socket);
```

```java
        }

        @Override
        public void dealwithMessage(String message) {
            if(message.contains("connect")) {
                String friendName = message.split(":")[0];
                JOptionPane.showMessageDialog(null,friendName + "连接成功");
            }
        }
    }

    public static void main(String[ ] args) {
        new Client().service();
    }
}
```

运行服务器端程序,启动服务器。然后再运行两次客户端,分别在输入框输入用户名,单击"确定"按钮,程序会显示客户端连接成功。

完成了客户端与客户端之间进行通信的程序之后,接下来完成双人版的"飞机大战"游戏。客户端需要实现双人联机版的"飞机大战"游戏,首先要重点实现的功能是:当按键控制"飞机"运动的时候,需要向另外一个客户端发送对应的信息,保证数据的同步。

重构单机版的 PlaneGame 类,在其中新增成员变量 planefriends 存储友机的数据信息,新增成员变量 client 与服务器端进行通信,代码如下:

```java
public class PlaneGame extends GameCore{
    private Screen screen;
    Plane plane;
    Plane planefriends;
    String name;
    String riendName;
    Client client;

    @Override
    public void initSprite() {
        final int SIZE = 8;
        name = JOptionPane.showInputDialog("请输入用户名:");
        if(name != null) {
            client = new Client();
            client.service();
            screen = new Screen(SIZE);
            plane = new Plane(7,4);
            planefriends = new Plane(7,4);
            screen.add(plane);
            screen.add(planefriends);
        }
    }

    @Override
    public void update() {
        char key = screen.getKey();
```

```java
            screen.delay();
            if(key == 'w') {
                plane.moveUp();
            }
            if(key == 's') {
                plane.moveDown();
            }
            if(key == 'a') {
                plane.moveLeft();
            }
            if(key == 'd') {
                plane.moveRight();
            }
        }
    }

    public class Client {
        private String host = "localhost";                    // 默认连接到本机
        private int port = 8080;                              // 默认连接到端口 8080
        private ClientThread thread;
        public Client() {
        }
        public Client(String host, int port) {                // 连接到指定的主机和端口
            this.host = host;
            this.port = port;
        }

        public void service(){
            try {
                Socket socket = new Socket(host,port);
                thread = new ClientThread(socket);
                String send = name + ":connect";
                sendMessage(send);
            }catch (IOException e) {
                e.printStackTrace();
            }
            new Thread(thread).start();
        }

        public void sendMessage(String message) {
            thread.sendMessage(message);
        }
        public class ClientThread extends ComThread{
            public ClientThread(Socket socket) {
                super(socket);
            }

            @Override
            public void dealwithMessage(String message) {
                if(message.contains("connect")) {
                    friendName = message.split(":")[0];
                    JOptionPane.showMessageDialog(null,friendName + "连接成功");
                }
            }
        }
    }
}
```

运行程序,登录成功后,出现"飞机大战"游戏界面。

接下来实现按键控制飞机运动的功能。与单机版本相比,除了在本机上响应按键行为外,还要将行为信息发送到另外一个客户端,代码如下:

```java
public class PlaneGame extends GameCore{
    private Screen screen;
    Plane plane;
    Plane planefriends;
    String name;
    String friendName;
    Client client;

    @Override
    public void initSprite() {
        final int SIZE = 8;
        name = JOptionPane.showInputDialog("请输入用户名:");
        if(name != null) {
            client = new Client();
            client.service();
            screen = new Screen(SIZE);
            plane = new Plane(7,4);
            planefriends = new Plane(7,4);
            screen.add(plane);
            screen.add(planefriends);
        }
    }

    @Override
    public void update() {
        char key = screen.getKey();
        screen.delay();
        if(key == 'w') {
            plane.moveUp();
            String send = name + ":" + friendName + ":" + "moveUp:Action";
            client.sendMessage(send);
        }

        if(key == 's') {
            plane.moveDown();
            String send = name + ":" + friendName + ":" + "moveDown: Action ";
            client.sendMessage(send);
        }

        if(key == 'a') {
            plane.moveLeft();
            String send = name + ":" + friendName + ":" + "moveLeft: Action";
            client.sendMessage(send);
        }

        if(key == 'd') {
            plane.moveRight();
            String send = name + ":" + friendName + ":" + "moveRight: Action";
```

```java
            client.sendMessage(send);
        }
}

public class Client {
    private String host = "localhost";           // 默认连接到本机
    private int port = 8080;                     // 默认连接到端口 8080
    private ClientThread thread;
    public Client() {
    }

    public Client(String host, int port) {       // 连接到指定的主机和端口
        this.host = host;
        this.port = port;
    }

    public void service(){
        try {
            Socket socket = new Socket(host,port);
            thread = new ClientThread(socket);
            String send = name + ":connect";
            sendMessage(send);
        }catch (IOException e) {
            e.printStackTrace();
        }
        new Thread(thread).start();
    }

    public void sendMessage(String message) {
        thread.sendMessage(message);
    }

    public class ClientThread extends ComThread{
        public ClientThread(Socket socket) {
            super(socket);
        }

        @Override
        public void dealwithMessage(String message) {
            if(message.contains("moveUp")) {
                planefriends.moveUp();
            }
            if(message.contains("moveDown")) {
                planefriends.moveDown();
            }
            if(message.contains("moveLeft")) {
                planefriends.moveLeft();
            }
            if(message.contains("moveRight")) {
                planefriends.moveRight();
            }

            if(message.contains("connect")) {
```

```
                    friendName = message.split(":")[0];
                    JOptionPane.showMessageDialog(null,friendName + "连接成功");
                }
            }
        }
    }
}
```

服务器端代码也要进行相应的修改,在信息处理方法 dealwithMessage()中增加相应的代码处理客户端发送的行为信息,代码如下:

```
public class Server {
    HashMap<String,ServerThread> hashmapThread;
    private int port = 8080;                    // 默认服务器端口
    public Server() {
        hashmapThread = new HashMap<String,ServerThread>();
    }

    public Server(int port) {                   // 创建指定端口的服务器
        this.port = port;
        hashmapThread = new HashMap<String,ServerThread>();
    }

    public void service(){
        try {
            ServerSocket server = new ServerSocket(port);
            Socket socket;
            System.out.println("服务器:启动,等待客户端的连接***");
            while(true) {                       // 等待客户连接
                socket = server.accept();
                ServerThread serverThread = new ServerThread(socket);
                new Thread(serverThread).start();
            }
        }catch (IOException e) {
            e.printStackTrace();
        }
    }
}
public class ServerThread extends ComThread{
    public ServerThread(Socket socket) {
        super(socket);
    }

    @Override
    public void dealwithMessage(String message) {
            if(message.contains("Action")) {
                String to = message.split(":")[1];
                String mess = message.split(":")[2];
                ServerThread thread = hashmapThread.get(to);
                thread.sendMessage(mess);
            }

            if(message.contains("connect")) {
```

```
                    String name = message.split(":")[0];
                    hashmapThread.put(name, this);
                    for(String key:hashmapThread.keySet()) {
                        ServerThread thread = hashmapThread.get(key);
                        if(thread != this) {
                            thread.sendMessage(message);
                            this.sendMessage(key + ":connect");
                        }
                    }
                }
            }
        }
    }

    public static void main(String[ ] args) {
        new Server(8080).service();
    }
}
```

运行服务器端程序,启动服务器,然后运行两次客户端程序,分别输入用户名之后,建立连接。在其中一个客户端通过按键控制飞机运动,在另外一个客户端可以观察到飞机也进行了同步运动。

本案例只实现了两个客户端之间进行联机游戏,读者可以尝试修改程序,实现多个客户端之间进行联机游戏。

在完成网络版"飞机大战"游戏之后,读者可以尝试建立网络游戏的框架类,以便设计新的网络游戏时可以对代码进行复用。

习题

14.1 设计自己感兴趣的网络版小游戏。

参 考 文 献

[1] ECKLE B.Java编程思想[M].陈昊鹏,译.4版.北京:机械工业出版社,2007.
[2] BRACKEEN D.Java游戏编程[M].邱仲潘,译.北京:科学出版社,2004.
[3] FOWLER M.重构改善既有代码的设计[M].熊节,林从羽,译.2版.北京:人民邮电出版社,2015.
[4] HORSTMANN C S.Java核心技术·卷1[M].林琪,苏钰涵,译.12版.北京:人民邮电出版社,2022.
[5] 黑马程序员.Java基础入门[M].2版.北京:清华大学出版社,2018.

图书资源支持

感谢您一直以来对清华版图书的支持和爱护。为了配合本书的使用,本书提供配套的资源,有需求的读者请扫描下方的"书圈"微信公众号二维码,在图书专区下载,也可以拨打电话或发送电子邮件咨询。

如果您在使用本书的过程中遇到了什么问题,或者有相关图书出版计划,也请您发邮件告诉我们,以便我们更好地为您服务。

我们的联系方式:

地　　址:北京市海淀区双清路学研大厦A座714

邮　　编:100084

电　　话:010-83470236　010-83470237

客服邮箱:2301891038@qq.com

QQ:2301891038(请写明您的单位和姓名)

资源下载:关注公众号"书圈"下载配套资源。

书圈

清华计算机学堂

观看课程直播